Two Manuals Combined into One Book

Browning Machine Gun .50 HB M2

and

40-MM Grenade Launcher M203

Department of Defense

Pentagon
Publishing

*** C1, FM 23-65**

FIELD MANUAL
No. 23-65

HEADQUARTERS
DEPARTMENT OF THE ARMY
Washington, DC, 19 June 1991

BROWNING MACHINE GUN CALIBER .50 HB, M2

CONTENTS

DISTRIBUTION RESTRICTION: Approved for public release; distribution is unlimited.

---

*This publication supersedes FM 23-65, 19 May 1972, and TC 23-65-1, 19 September 1984.

* PREFACE

This manual provides technical information, training techniques, and guidance on the caliber .50 HB machine gun, M2. Unit leaders and designated gunners will use this information to successfully integrate the weapon into combat operations. They can instruct on the range or at concurrent training stations.

The material applies as is to both nuclear and conventional warfare.

Trainers must ensure that everyone observes safety procedures at all times. Commanders, trainers, and individual students must remember that safety is everyone's responsibility. Leaders will conduct all training as though each weapon were fully loaded. At no time during training will anyone allow the desire for speed or accuracy to override the requirement to follow safety procedures. Safe training is good training.

The proponent of this publication is US Army Infantry School. Send comments and recommendations on DA Form 2028 (Recommended Changes to Publications and Blank Forms) directly to the Commandant, US Army Infantry School, ATTN: ATSH-INB-O, Fort Benning, GA 31905-5594 or lusanoh@benning.army.mil.

Unless otherwise stated, the masculine gender refers to both men and women.

CHAPTER 1

INTRODUCTION

The procedures and methods used in Army machine gun marksmanship are based on the concept that soldiers must be skilled gunners who can effectively apply their firing skills in combat. The basic firing skills and exercises outlined in this manual must be a part of every unit's machine gun training program. The soldiers' proficiency depends on proper training and application of basic gunnery fundamental, which are taught in a progressive program to prepare gunners for combat.

1-1. TRAINING STRATEGY

Training strategy is the overall concept for integrating resources into a program to train individual and collective skills needed to perform a unit's wartime mission.

a. Training strategies for marksmanship are implemented in TRADOC institutions (NCOES, basic and advanced officer's courses) and in units. The overall training strategy is multifaceted and is inclusive of the specific strategies used in institution and unit programs. Also included are the supporting strategies that use resources such as publications, ranges, ammunition, training aids, devices, simulators, and simulations. These strategies focus on developing critical soldier skills, and on leader skills that are required for the intended outcome.

b. Two primary components compose the training strategies: initial training and sustainment training. Both may include individual and collective skills. Initial training is critical because a task that is taught correctly and learned well is retained longer. Well-trained skills can be more quickly regained and sustained if an interim of nonuse occurs. The more difficult and complex the task, the harder it is to sustain the skill. Personnel turnover is a main factor in decay of collective skills, since the loss of critical team members requires retraining to regain proficiency. If a long period elapses between initial and sustainment training sessions or training doctrine is altered, retraining maybe required.

c. The training strategy for caliber .50 MG marksmanship begins in selected resident training and continues in the unit. An example of this overall process is illustrated in Figure 1-1 and provides a concept of the flow of unit sustainment training. The soldiers graduating from selected resident training courses have been trained to maintain their MGs and to hit a variety of targets. They have learned range determination, target detection, application of marksmanship fundamentals, and other skills needed to engage a target. Task training during these courses may lead to qualification.

d. Training continues in units on the basic skills taught in combat arms. Additional skills, such as suppressive fire and supporting fire, are trained and then integrated into collective training exercises, which include platoon and squad live-fire STXs. (A unit-marksmanship training program is explained in Chapter 5.) The strategy for sustaining the basic marksmanship skills taught in combat arms is periodic preliminary instruction, followed by qualification range firing. However, a unit must set up a year-round program to sustain skills. Key elements include training of trainers and refresher training of nonfiring skills.

e. Additional skills trained in the unit include techniques for employment, suppressive fires, night fire, MOPP firing, and moving targets. Related soldier skills of camouflage, cover and concealment, maneuver, and preparation and selection of a fighting position are addressed in STP 21-24-SMCT, which must be integrated into tactical training.

f. In the unit, individual and leader proficiency of marksmanship tasks are integrated into collective training to include squad, section, and platoon drills and STXs. The collective tasks in these exercises, and how they are planned and conducted, are in the MTP and battle drill books for each organization. Based on the type organization, collective tasks are evaluated to standard and discussed during leader and trainer after-action reviews. Objective evaluations of both individual and unit proficiency provide readiness indicators and future training requirements.

g. A critical step in the Army's overall marksmanship training strategy is to train the trainers and leaders first. Leader courses and unit publications develop officer and NCO proficiencies necessary to plan and conduct marksmanship training and to evaluate the effectiveness of unit marksmanship programs. Training support materials are provided by the proponent schools to include field manuals, training aids, devices, simulators, and programs that are doctrinal foundations and guidance for training the force.

h. Once the soldier understands the weapon and has demonstrated skill in zeroing, additional live-fire training and a target acquisition exercise at various ranges are conducted. Target types and scenarios of increasing difficulty must be mastered to develop proficiency.

i. Initial individual training culminates in the soldier's proficiency assessment, which is conducted on a transition/record fire range. This evaluation also provides an overview of unit proficiency and training effectiveness.

j. Unit training programs maintain the soldiers' proficiency level. The ultimate goal of a unit marksmanship program is to maintain well-trained gunners so a unit can survive and win on the battlefield. The trainer must realize that qualification is not an end, but a step toward reaching this combat requirement. (See Figure 1-1.)

(1) To reach this goal, the gunner must be able to position and use his weapon under the following combat conditions:

- Enemy personnel are seldom visible except when assaulting.
- Most combat fire must be directed at an area where the enemy has been detected or where he is suspected of being located but cannot be seen. Area targets consist of objects or outlines of men irregularly spaced along covered and concealed areas (ground folds, hedges, or borders of woods).
- Most combat targets can be detected by smoke, flash, dust, noise, or movement and are visible only for a moment.
- Some combat targets can be engaged by using nearby objects as reference points.
- The nature of the target and irregularities of terrain and vegetation may require a firer to use a variety of positions in addition to the prone or supported position to fire effectively on the target. In a defensive situation, the firer usually fires from a supported position.
- Most combat targets have a low contrast outline and are obscure. Therefore, choosing an aiming point in elevation is difficult.
- Time-stressed fire in combat can be divided into three types: a single, fleeing target that must be engaged quickly; distributed targets engaged within the time they remain available; and a surprise target that must be engaged at once with accurate, instinctive fire.

(2) The unit's program must provide fundamental training to sustain and improve the skills and proficiency the soldier has attained during his basic marksmanship training. Once basic skills have been mastered, these must be improved by conducting new or advanced individual and collective training. The program must develop collective firing skills by incorporating marksmanship into tactical exercises. This training must maintain the soldier's confidence in the weapon and his skills. A soldier's

survival may depend on his ability to defend himself or other members of the unit. Therefore, individual and collective firing skills must support the expected battlefield conditions and the unit's combat mission.

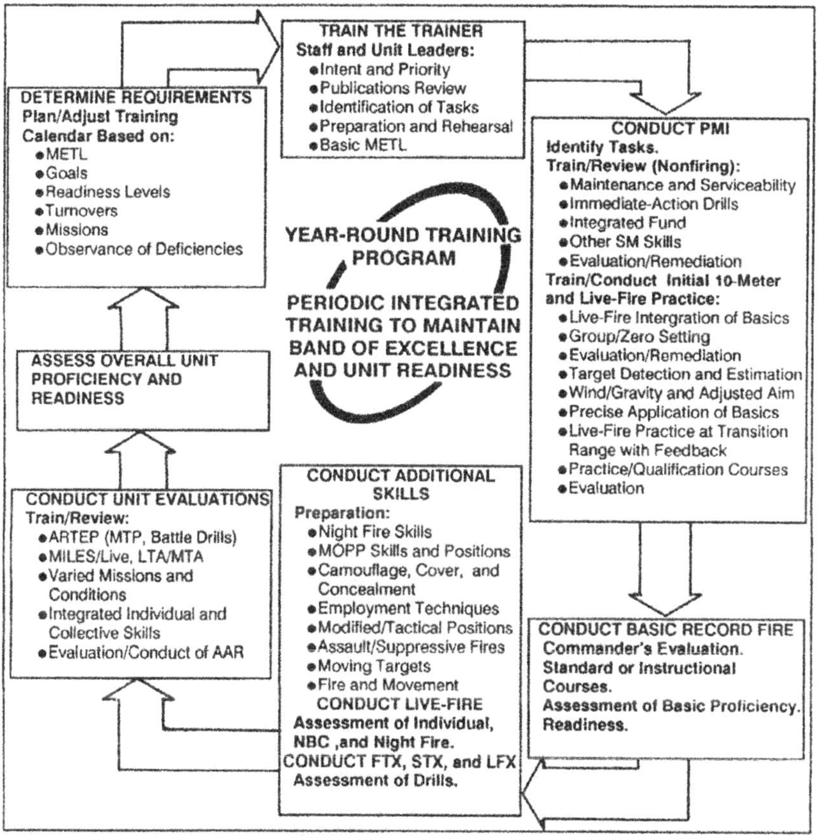

Figure 1-1. Unit marksmanship sustainment strategy.

(3) A unit's marksmanship program must be battlefield oriented. It must be based upon several individual combat tasks as well as organizational, operational, or contingency missions. It must have available resources such as ammunition, time, ranges, and qualified trainers. This manual provides the information a unit commander needs to develop an effective marksmanship program for his unit requirements.

(4) General marksmanship, training knowledge, and accurate firing are acquired skills that perish easily. Skill practice should be conducted for short periods throughout the year. Most units have a readiness requirement that all soldiers must zero their MGs within a certain time after unit assignment. Also, soldiers must confirm the zero of their assigned MGs before conducting a qualification firing.

1-2. DESCRIPTION

The Browning machine gun caliber .50 HB, M2 (Figure 1-2) is a belt-fed, recoil-operated, air-cooled, crew-served machine gun. The gun is capable of single shot, as well as automatic fire, and operates on the short recoil principle.

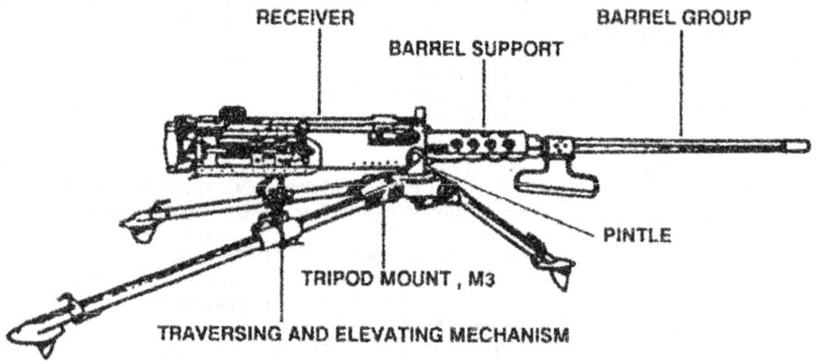

Figure 1-2. Browning machine gun.

a. The machine gun is capable of being fed from either the right or left by repositioning certain parts. The weapon has nonfixed headspace that must be set. Timing must also be adjusted to cause the gun to fire slightly out of battery to prevent damage to moving parts. The force for recoil operation is furnished by expanding powder gases, which are controlled by various springs, cams, and levers. Maximum surface of the barrel and receiver are exposed to permit air cooling. Perforations in the barrel support allow air to circulate around the breach end of the barrel and help in cooling the parts. A heavy barrel is used to retard early overheating.

b. The gun has a leaf-type rear sight (Figure 1-3), graduated in both yards and roils. The scale ranges from 100 to 2,600 in yards, and from 0 to 62 in mils. The windage knob-permits deflection changes to right or left of center. The front sight is a fixed blade type with cover (Figure 1-4).

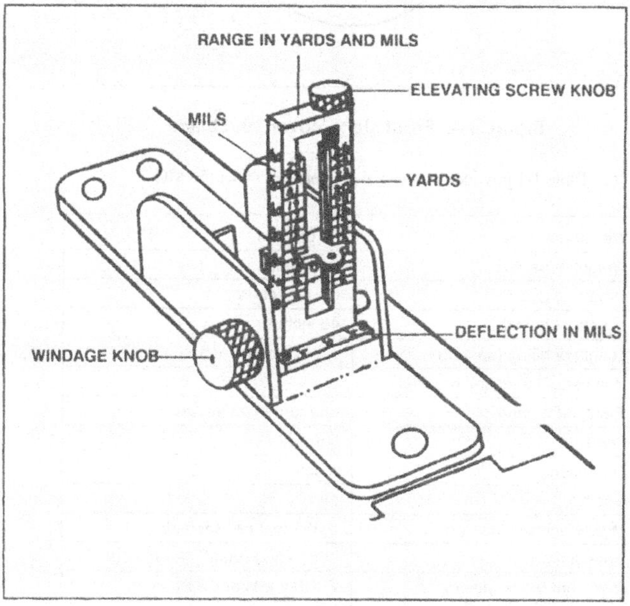

Figure 1-3. Leaf type rear sight.

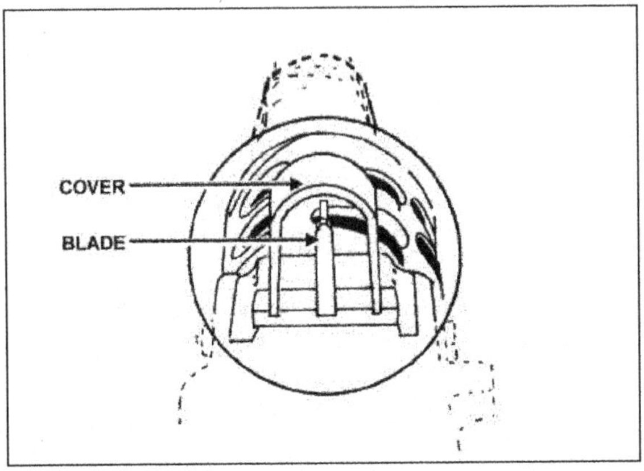

Figure 1-4. Front sight, cover, and blade.

c. Table 1-1 provides the general data on the caliber .50 MG.

Weight (approx)	84 pounds
Weight of barrel	24 pounds
Length of gun	65.13 inches
Length of barrel	45 inches
Length of rifling (approx)	41.88 inches
Number of lands and grooves	8
Twist, right-hand	one turn in 15 inches
Feed	link-belt
Operation	recoil
Cooling	air
Muzzle velocity (approx)	3,050 feet per second
Rate of fire (cyclic)	450 to 550 rounds per minute
Maximum range (approx)	7,440 yards or 6,764 meters
Maximum effective range (approx)	2,000 yards or 1,830 meters
• Area targets	1,830 meters
• Point targets, single shot	1,500 meters

Table 1-1. General data.

1-3. COMPONENTS

The major components of the caliber .50 MG and their purposes are shown in Table 1-2 and Figure 1-5.

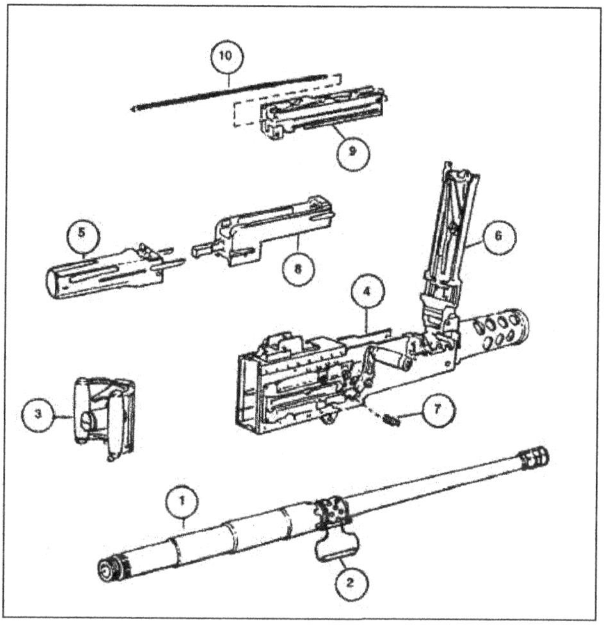

Figure 1-5. Components of the caliber .50 MG.

COMPONENTS	PURPOSES
l. Barrel Group	Houses cartridges for firing; directs projectile.
2. Carrier Assembly	Provides handle to carry barrel and to remove the barrel from the receiver.
3. Backplate Group	Houses the trigger, bolt latch release, buffer tube sleeve, and the left and right spade grips.
4. Receiver Group	Serves as a support for all major components; houses action of weapon, which controls functioning of weapon.
5. Bolt Group	Provides feeding, chambering, firing, and extracting, using the propellant gases and recoil spring for power.
6. Cover Group	Feeds linked belt ammunition; positions and holds cartridges in position for extracting, feeding, and chambering.
7. Bolt Stud	Provides a means to move the bolt to the rear with the retracting slide handle.
8. Barrel Extension Group	Secures the barrel to the recoiling parts.
9. Barrel Buffer Body	Assists in recoil and counterrecoil of the bolt group.
10. Driving Spring Rod Assembly	Drives the bolt forward when the bolt latch release is depressed.

Table 1-2. Components and their purposes.

1-4. GROUND MOUNTS

The two principal ground mounts used with the caliber .50 machine gun are the tripod mount, M3, and the antiaircraft mount, M63. The tripod mount, M3, is a ground mount designed for use against ground targets. The antiaircraft mount,

M63, is a ground mount principally designed for use against aerial targets. Its use against ground targets is limited because the mount tends to be unstable when the gun is fired at low angles.

a. **Tripod Mount, M3**. The M3 mount is the standard ground mount of the caliber .50 machine gun (Figure 1-6). It is a folding tripod with three, telescopic, tubular legs connected at the tripod head. Each leg ends in a metal shoe that can be stamped into the ground for greater stability. The two trail legs are joined together by the traversing bar. The traversing bar serves as a support for the traversing and elevating mechanism, which in turn supports the rear of the gun. The tripod head furnishes a front support for the mounted gun that is further supported by the short front leg. When the tripod is emplaced on flat terrain with all extensions closed, the adjustable front leg should form an angle of about 60 degrees with the ground. This places the gun on a low mount about 12 inches above the ground. To raise the tripod farther off the ground, extend the telescopic front and trail legs enough to keep the tripod level and maintain the stability of the mount.

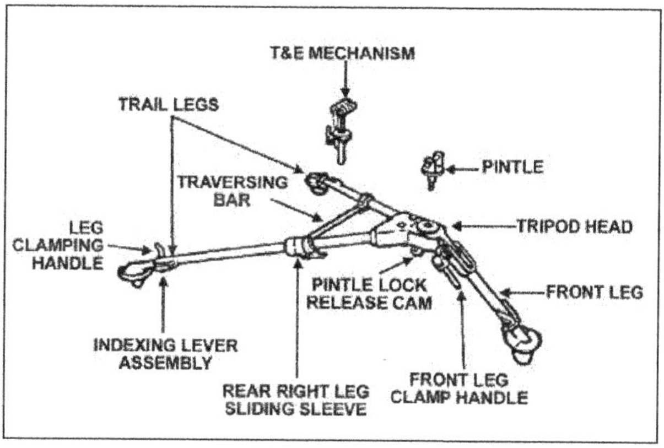

Figure 1-6. M3 tripod mount.

(1) To set the tripod trail legs--

(a) Unscrew the leg-clamping handle, press down on the indexing lever, and extend the leg to the desired length.

(b) Align the indexing lever stud with one of the holes in the tripod leg extension.

(c) Release the pressure on the indexing lever, allowing the stud to fit the desired hole. Tighten the leg-clamping handle.

(2) *To set the front leg of the tripod-*

(a) Turn the front leg clamp handle counterclockwise to loosen the front leg.

(b) Adjust the leg to the desired angle and tighten the front leg clamp.

(3) To secure the tripod legs, stamp the metal shoe on each tripod leg into the ground. Sandbag each leg to stabilize the M2 for firing.

b. **Antiaircraft Mount, M63.** The antiaircraft mount (Figure 1-7) is a four-legged, low silhouette, portable mount used for antiaircraft fire. Table 1-3 lists the general data pertaining to the M63.

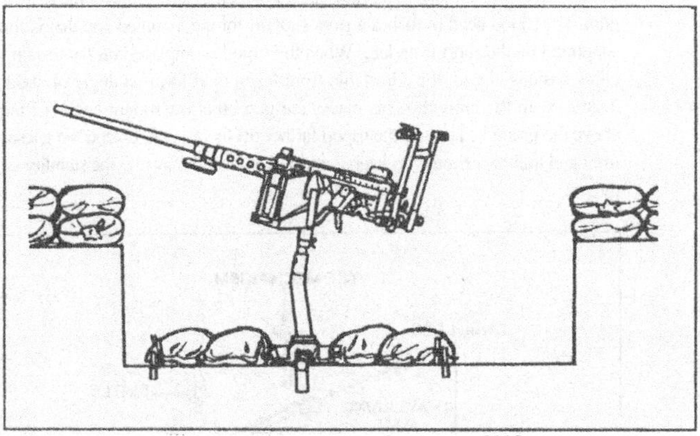

Figure 1-7. Antiaircraft mount, M63.

WEIGHTS:	
Overall	144 pounds
Four legs	24 pounds
Base assembly	54 pounds
Elevator assembly	12 pounds
Cradle assembly	44 pounds
Ammunition box-tray assembly	10 pounds
HEIGHT OVERALL	42 inches
LENGTH OF LEG	24 inches
DIAMETER OF BASE (WITH LEG ASSEMBLED)	52 inches
MAXIMUM ELEVATION	85 degrees
MAXIMUM DEPRESSION	29 degrees
MAXIMUM TRAVERSE	360 degrees

Table 1-3. M63 general data.

1-5. ACCESSORIES FOR GROUND MOUNTS

The following paragraph explains the functions of the traversing and elevating mechanism and pintle used in the mounting of the machine gun when used in the ground configuration.

a. **Traversing and Elevating Mechanism.** The T&E mechanism (Figure 1-8) is used to engage preselected target areas at night or during limited visibility conditions. Record direction and elevation readings from the traversing bar and T&E mechanism. Record all readings in mils.

(1) The traversing mechanism consists of a traversing bar, slide, and screw assembly.

(a) The traversing bar, graduated in 5-roil increments, fits between the trail legs of the tripod. The traversing slide and screw assembly are clamped in place on the traversing bar by the traversing slide lock lever. When the traversing slide is locked to the traversing bar, the traversing handwheel should be centered. The traversing slide is properly mounted when the lock lever is to the rear and the traversing handwheel is positioned to the left.

(b) To make changes in direction, loosen the traversing slide lock lever and move the slide along the traversing bar. This permits traverse of 400 mils left or right of the zero index in the center of the traversing bar. Readings on the traversing bar are taken from the left side of the traversing slide. For changes of 50 mils or less in deflection, turn the traversing handwheel of the screw assembly. This allows a traverse of 50 mils left or right of center. One click in the traversing handwheel signifies 1 mil change in direction.

(2) The elevating mechanism consists of an upper and lower elevating screw.

(a) It is connected to the gun by inserting the quick release pin assembly through the holes in the upper elevating screw yoke and the rear mounting lugs of the receiver. A scale, graduated in mils, is fitted to the upper screw to indicate elevation. This scale is marked to show 250 mils in depression and 100 mils in elevation from the zero setting.

(b) The elevating handwheel is graduated in 1-mil increments up to 50 mils, and is fastened to the elevating screw by a screw lock. This synchronizes the handwheel graduations with those on the upper elevating screw. A spring-actuated index device produces a clicking sound when the handwheel is turned. Each click equals 1 mil change in elevation. The handwheel is turned clockwise to depress the barrel and counterclockwise to elevate.

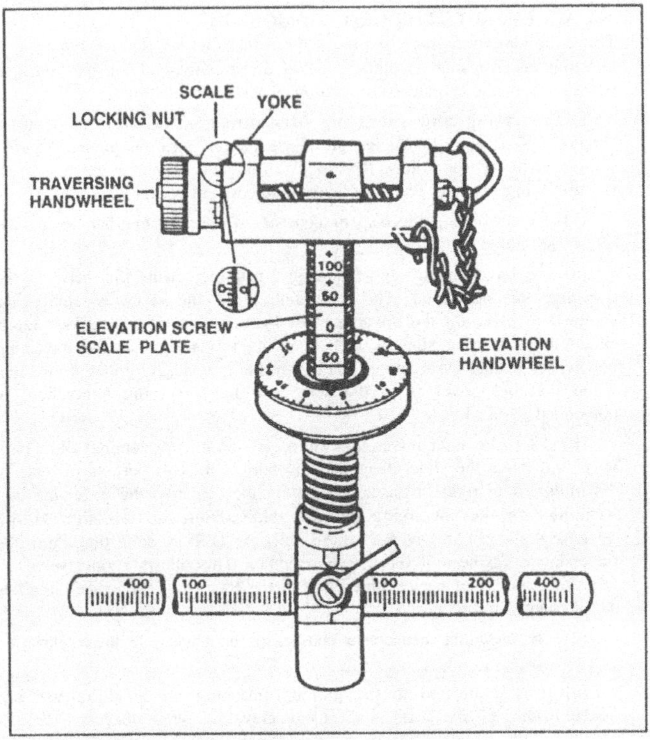

Figure 1-8. Traversing and elevating mechanism.

b. **Pintle.** The gun is connected to the tripod mount, M3, by a pintle (Figure 1-9, page 1-14). This pintle is semipermanently attached to the machine gun by a pintle bolt through the front mounting hole in the receiver. The tapered stem of the pintle seats in the tripod head. It is held secure by a pintle lock and spring. To release the pintle, raise the pintle lock, releasing the cam. The weight of the pintle and traversing and elevating mechanism are considered as part of the total weight of the tripod mount, M3 (44 pounds).

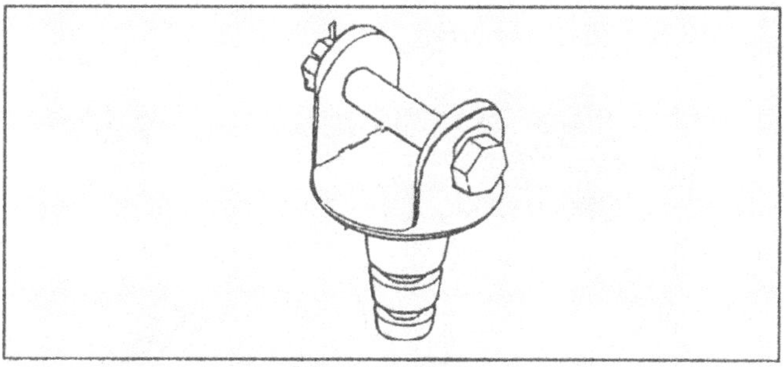

Figure 1-9. Pintle.

1-6. VEHICULAR MOUNTS

The four principal vehicular mounts used with the caliber .50 machine gun are the truck mount, M36; the pedestal truck mount, M31C and M24A2; the commander's cupola, Ml13 armored personnel carrier; and the MK64 gun cradle.

a. **Truck Mount, M36.** This mount consists of a cradle with a roller carriage on a circular track (Figure 1-10). The cradle can be rotated in the pintle sleeve of the carriage and can be adjusted for elevation. The carriage is guided on the track by rollers. The track is secured to the vehicle by supports.

(1) To move the gun in elevation on the M36 mount, remove the cradle locking pin and place it in the carriage handle; grasp the spade grips and elevate or depress as desired. The gun is also moved in traverse by pressure on the spade grips.

(2) To move the gun on the track, raise the brake handle lever until it is retained by the brake detent plungers. The cradle may then be moved on the track by applying pressure on the carriage handle.

b. **Pedestal Truck Mount, M31C.** Pedestal mounts are component assemblies designed for installation on the 1/4-ton vehicles to support a machine gun mount. They are composed of a pintle socket, pintle clamping screw column, and braces (Figure 1-11).

c. **Armored Vehicle Cupola Mount.** A caliber .50 machine gun and mount are installed in the gun support on the commander's cupola of an Ml 13 armored personnel carrier. The machine gun can be traversed 360 degrees, elevated 53 degrees, and depressed 21 degrees maximum (Figure 1-12).

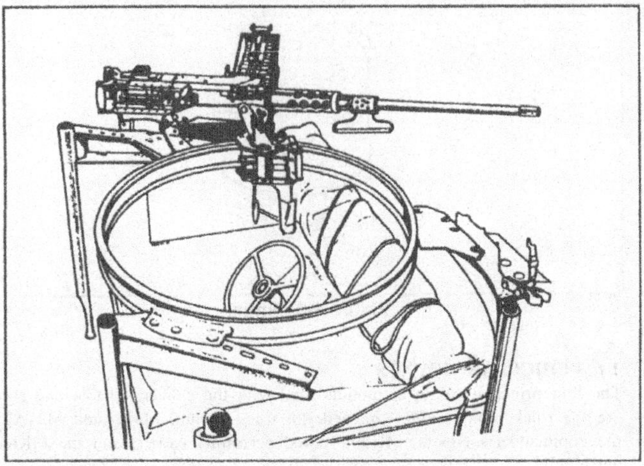

Figure 1-10. Truck mount, M36.

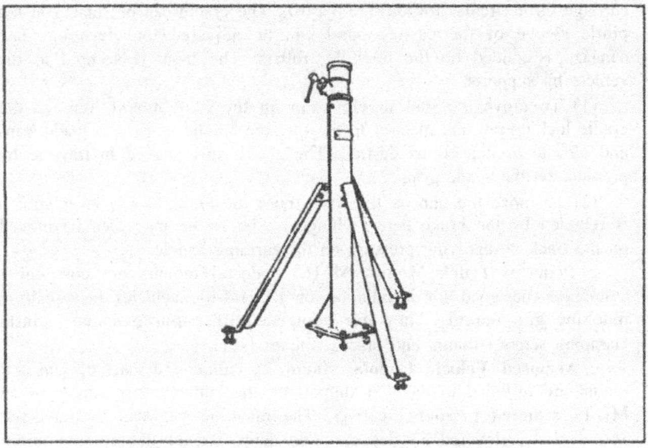

Figure 1-11. Pedestal truck mount, M31 C.

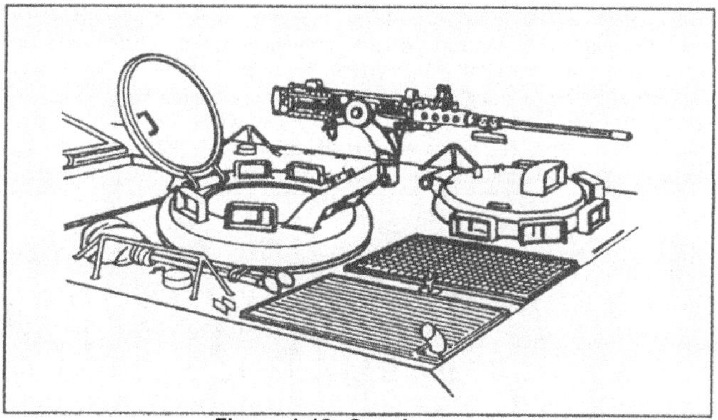

Figure 1-12. Cupola mount.

d. **MK64 Gun Cradle Mount.** This vehicle mount was primarily designed for the M2. However, because of its versatility, the MK64 will accept the MK 19 also (using the M2 mounting adapter assembly). The MK64 can be mounted on the following vehicles – M151 series, M966 HMMWV armament carrier, and the M113 series (Figure 1-13).

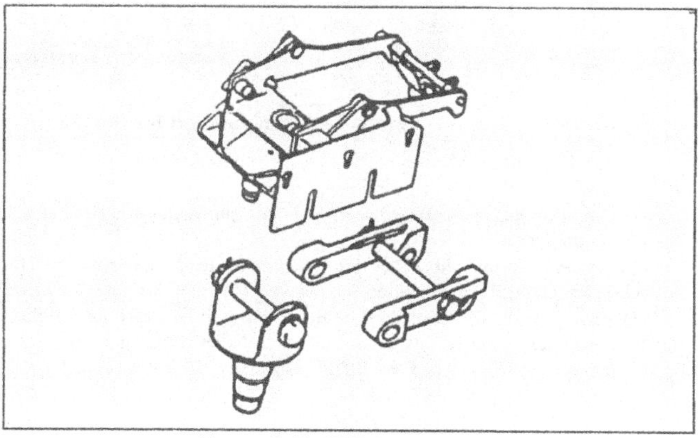

Figure 1-13. MK64 gun cradle.

1-7. AMMUNITION

This paragraph describes the ammunition used in caliber .50 machine guns (Figure 1-14). Soldiers should be able to recognize the types of ammunition available and know how to care for it. The caliber .50 cartridge consists of a cartridge case, primer, propelling charge, and the bullet. See TM 9-1300-200. The term bullet refers only to the small-arms projectile. There are eight types of ammunition issued for use in the caliber .50 machine gun. The tips of the various rounds are color-coded to indicate their type. The ammunition is linked with the M2 or M9 metallic links for use in the machine gun (Figure 1-15, page 1-18).

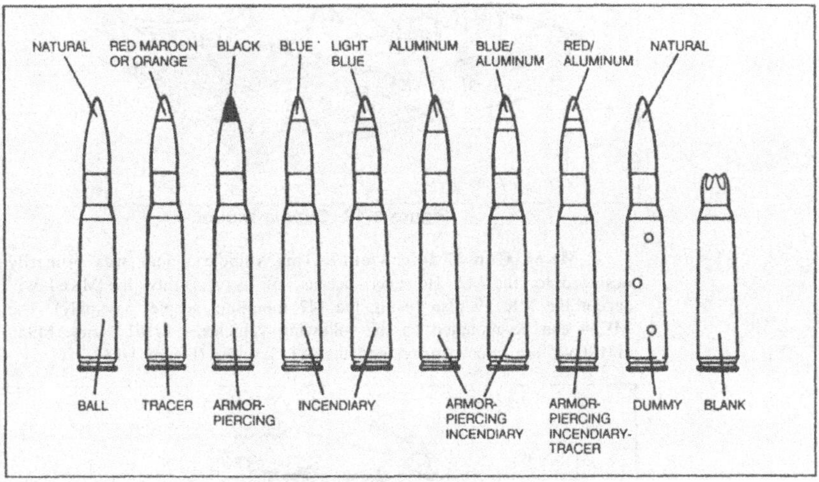

Figure 1-14. Ammunition for the M2.

a. **Classification.** The eight types of ammunition are used for the following purposes.

(1) *Ball.* For use in marksmanship training, and against personnel and light material targets.

(2) *Tracer.* To aid in observing fire. Secondary purposes are for incendiary effect and for signaling.

(3) *Armor-piercing.* For use against armored aircraft and lightly armored vehicles, concrete shelters, and other bullet-resisting targets.

(4) *Incendiary.* For incendiary effect, especially against aircraft.

(5) *Armor-piercing-incendiary.* For combined armor-piercing and incendiary effect.

(6) *Armor-piercing-incendiary-tracer.* For combined armor-piercing and incendiary effect, with the additional tracer feature.

(7) *Blank.* For simulated fire (contains no bullet).

(8) *High-pressure test.* For use only in proof firing of weapons and barrels.

(9) *Dummy.* For training (completely inert).

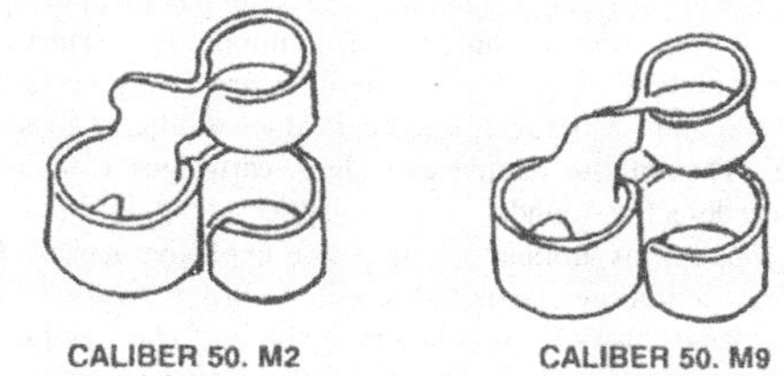

Figure 1-15. M2 and M9 links (closed loop).

b. **Ballistic Data.** The approximate maximum range and average muzzle velocity of some of the different types of caliber .50 ammunition authorized for use in the machine gun are noted in Table 1-4.

CARTRIDGE	MAXIMUM RANGE (meters)	METER TRACE	AVERAGE MUZZLE VELOCITY (feet per second)
Ball, M2	7,400	—	2,930
Tracer, M1 (with gilding metal jacket)	5,575	1,800	2,860
Tracer, M1 (with clad steel jacket)	5,450	1,800	3,030
Tracer, M17	5,450	2,450	3,030
Incendiary, M1	6,050	—	3,090
Armor-piercing, M2	7,400	—	2,930
Armor-piercing-incendiary,M8	6,470	—	3,050
Armor-piercing-incendiary-tracer, M20	6,470	*300-1,750	3,050

* This tracer is dim at near ranges but increases to bright as it moves farther from the gun.

Table 1-4. Ballistic data.

c. **Care, Handling, and Preservation**. Exercise care to prevent ammunition boxes from becoming broken or damaged. If they do, repair them immediately. Transfer all original markings to the new parts of the box. Do not open ammunition boxes until the ammunition is to be used. Ammunition removed from the airtight container, particularly in damp climates, is likely to corrode. Protect the ammunition from mud, sand, and water. If the ammunition gets wet or dirty, wipe it off at once with a clean, dry cloth. Wipe off light corrosion as soon as it is discovered. Turn in heavily corroded cartridges. Do not expose ammunition to the direct rays of the sun. If the powder is hot, excessive pressure may be developed when the weapon is fired. Do not oil or grease ammunition. Dust and other abrasives that collect on greasy ammunition are injurious to the operating parts of the gun. Moreover, oiled cartridges produce excessive chamber pressure. Do not fire dented cartridges, cartridges with loose bullets, or otherwise defective rounds.

d. **Storage**. Small-arms ammunition is not an explosive hazard, but under poor storage conditions it may become a fire hazard. Store ammunition of all classes away from radiators, hot water pipes, and other sources of heat. Whenever possible, store ammunition under cover. If it is necessary to leave ammunition in the open, keep it at least 6 inches off the ground and covered with a double thickness of tarpaulin. Place the tarpaulin so that it gives maximum protection and allows free circulation of air. Dig suitable trenches to prevent water from flowing under the ammunition pile.

e. **Miscellaneous Data**. Table 1-5 lists the maximum penetration in inches for an armor-piercing cartridge fired from the 45-inch barrel (muzzle velocity, 2,935 feet per second), which in some cases may enhance the leader's selection of targets to engage.

	INCHES AT:		
MATERIAL	**200 METERS**	**600 METERS**	**1,500 METERS**
Armor plate (homogeneous)	1.0	0.7	0.3
Armor plate (face-hardened)	0.9	0.5	0.2
Sand (100 pounds dry weight/cubic feet)	14.0	12.0	16.0
Clay (100 pounds dry weight/cubic feet)	28.0	27.0	21.0

Table 1-5. Maximum penetration for armor-piercing cartridge.

Table 1-6 lists the maximum penetration in inches for a ball cartridge fired from the 45-inch barrel (muzzle velocity, 2,935 feet per second):

	INCHES AT:		
MATERIAL	**200 METERS**	**600 METERS**	**1,500 METERS**
Sand (100 pounds dry weight/cubic feet)	14.0	12.0	6.0
Clay (100 pounds dry weight/cubic feet)	28.0	27.0	21.0
Concrete	2.0	1.0	1.0

Table 1-6. Maximum penetration for ball cartridge.

CHAPTER 2

MAINTENANCE

This chapter addresses the proper care of the machine gun to ensure its overall effectiveness and efficient functioning. The information includes the gunner's knowledge in disassembly and assembly, inspection, and maintenance procedures of the gun, its mount, the T&E, and its ammunition.

2-1. SAFETY

The paramount consideration while training with the machine gun is safety. It is imperative that the weapon be cleared properly before disassembly and inspection. (See Figure 2-1 for step-by-step procedures.)

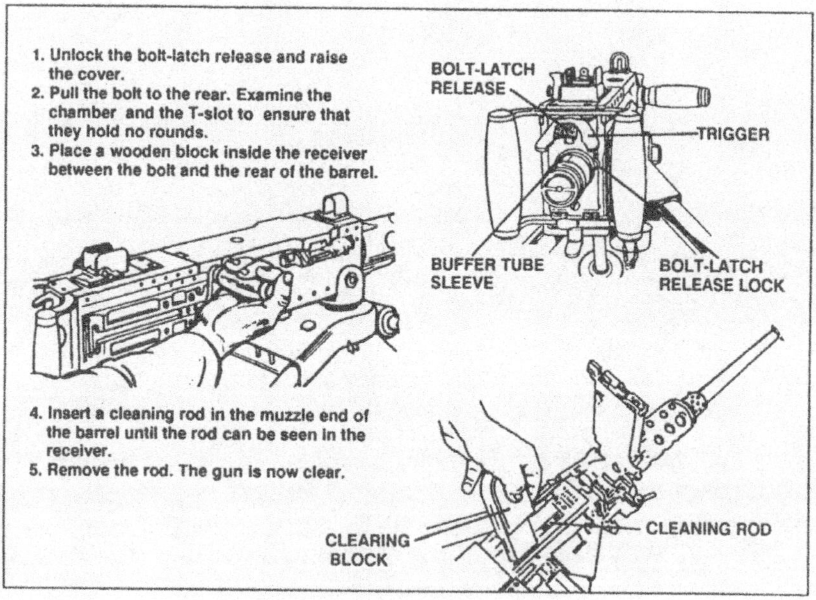

Figure 2-1. Clearing the weapon.

2-2. GENERAL DISASSEMBLY

The crew operating the MG must be fully familiar with its general disassembly, which consists of removing the major groups and assemblies for inspection and cleaning. The eight major groups that must be disassembled in the following order are barrel group, backplate group, driving spring rod assembly, bolt group, barrel extension group, barrel buffer body group, barrel buffer assembly, and receiver group (Figure 2-2).

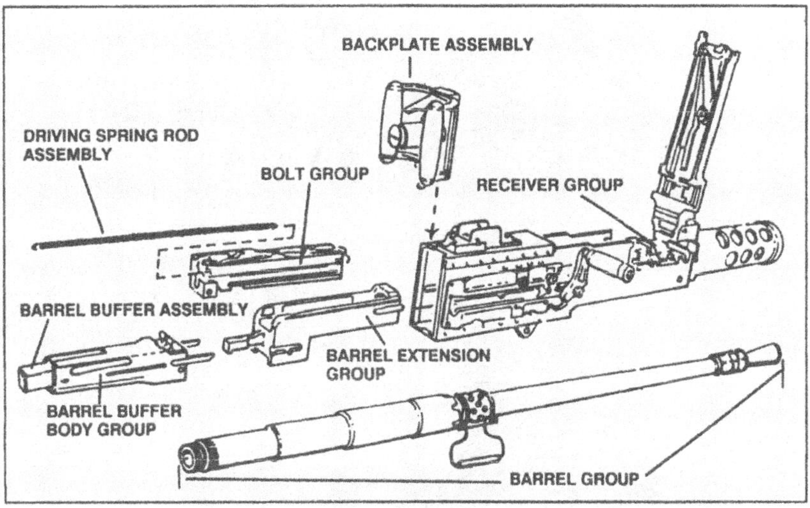

Figure 2-2. Major groups.

a. **Barrel Group.** Turn the cover latch and raise the cover group (Figure 2-3). Grasp the retracting slide handle with the right hand, palm up, and pull the recoiling parts to the rear until the lug on the barrel locking spring aligns with the 3/8-inch hole in the right sideplate of the receiver (just below the feedway exit). The barrel can be turned only when the lug is aligned with the 3/8-inch hole. Place the smallest loop of a caliber .50 link, or suitable spacer, between the trunnion block and the barrel extension (Figure 2-4). This holds the barrel locking spring lug aligned with the 3/8-inch hole in the right sideplate. Unscrew the barrel from the receiver (Figure 2-5, page 2-4). Be careful not to damage the threads or barrel locking notches when setting the barrel down. Pull back slightly on the retracting slide handle and remove the link or spacer from the receiver.

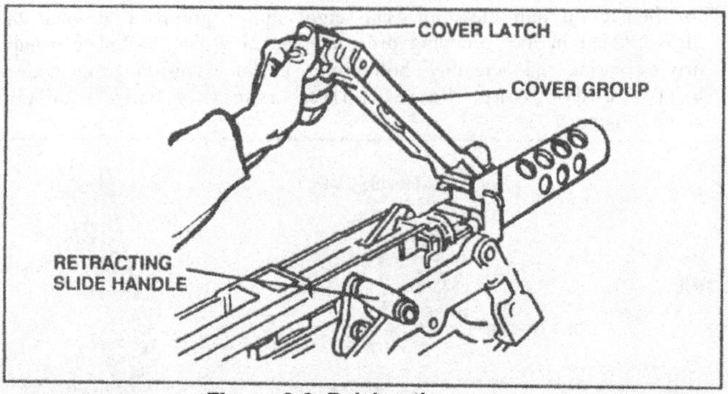

Figure 2-3. Raising the cover.

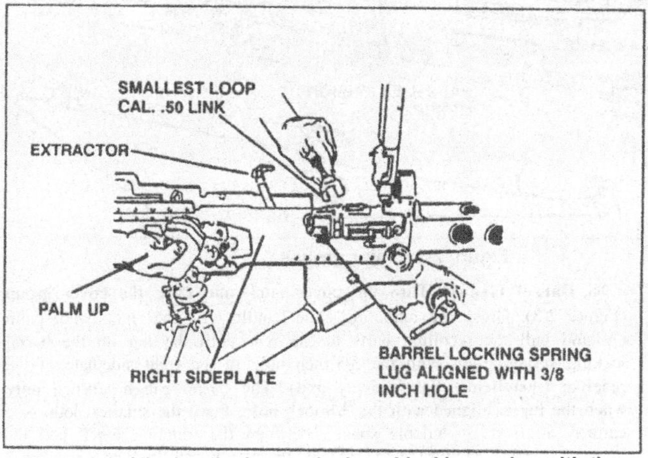

Figure 2-4. Aligning the lug on the barrel locking spring with the 3/8-inch hole in the right sideplate.

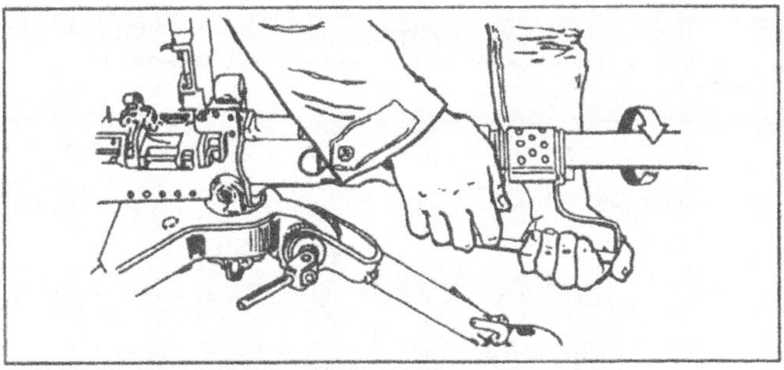

Figure 2-5. Removing the barrel.

b. **Backplate Group.** Ensure that the bolt latch release is up, free of the bolt latch release lock. If it is not, push down on the bolt latch release and turn the buffer tube sleeve to the right to free it (Figure 2-6). The bolt must be forward before the backplate is removed. If the bolt is to the rear, push down on the bolt latch release, place palm up on the retracting slide handle, and ease the bolt forward. The backplate latch lock and latch are below the buffer tube. Pull out on the lock and up on the latch; remove the backplate by lifting it straight up (Figure 2-7).

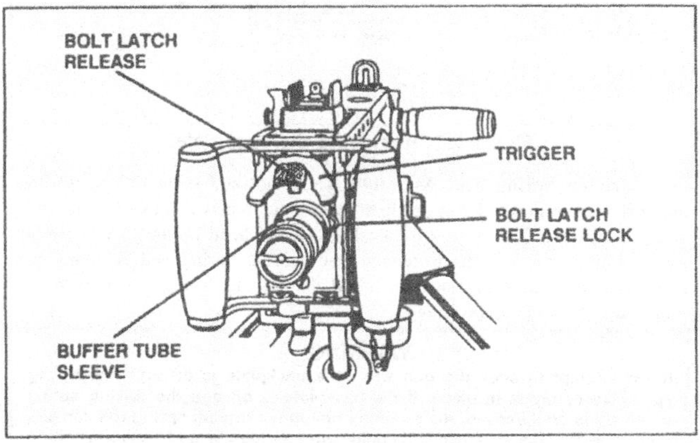

Figure 2-6. Releasing the bolt latch.

CAUTION

Care must be taken to prevent the bolt from slamming forward with the barrel removed.

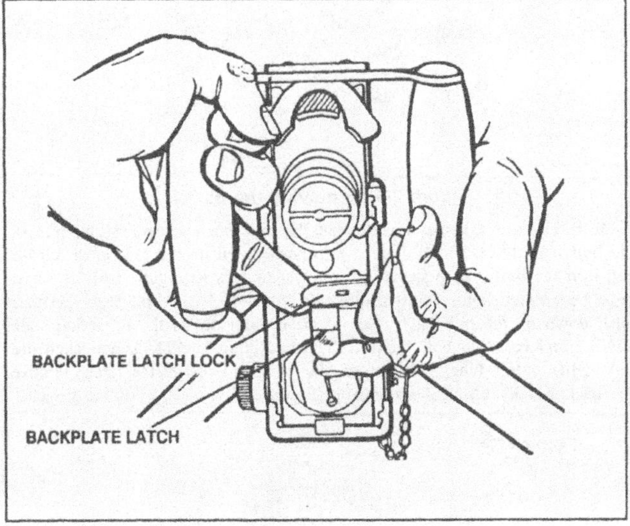

Figure 2-7. Removing the backplate.

c. **Driving Spring Rod Assembly.** The inner and outer driving springs and driving spring rod are located inside the receiver next to the right sideplate (Figure 2-8, page 2-6). Push in on the head of the driving spring rod and push to the left to remove the driving spring rod retaining pin from its seat in the right sideplate. Pull the driving spring assembly to the rear and out of the receiver.

WARNING

Never attempt to cock the gun while the backplate is off and the driving spring assembly is in place. If the backplate is off and the driving spring assembly is compressed, the retaining pin on the driving spring rod can slip from its seat in the sideplate and could cause serious injury to anyone behind the gun.

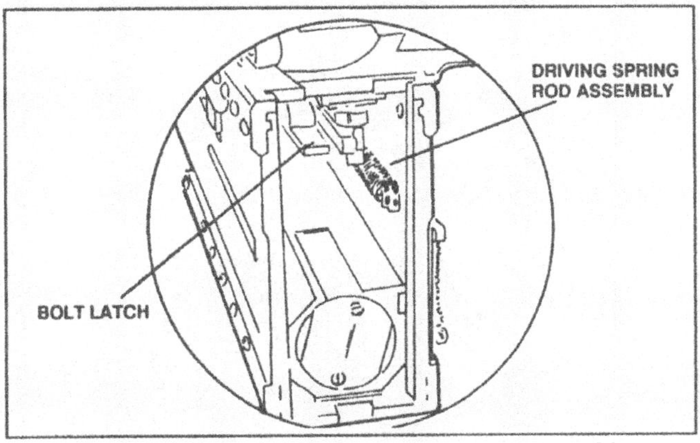

Figure 2-8. Removing the driving spring rod assembly.

d. **Bolt Stud.** Grasp the retracting slide handle and give it a quick jerk, freeing the bolt from the barrel extension. Align the collar of the bolt stud with the clearance hole in the bolt slot on the right sideplate, and remove the bolt stud (Figure 2-9). If the bolt is accidentally moved all the way to the rear, the bolt latch will engage in the bolt latch notches in the top of the bolt. If this occurs, raise the bolt latch (left of the trigger lever) and push the bolt forward to align the bolt stud with the clearance hole (Figure 2-10).

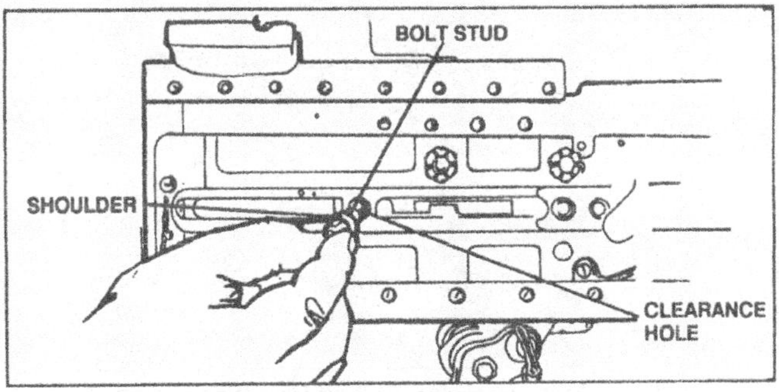

Figure 2-9. Removing the bolt stud.

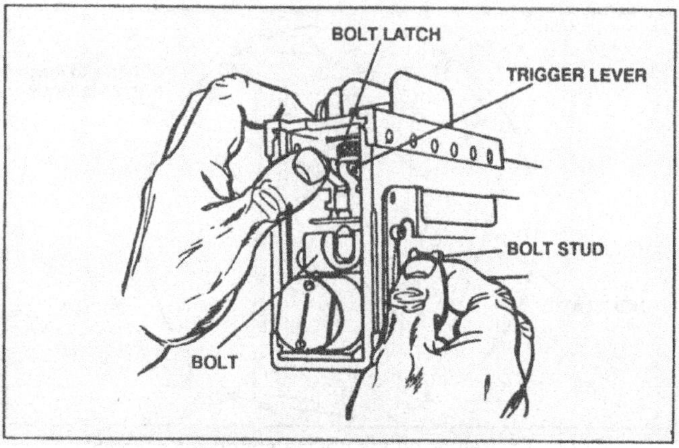

Figure 2-10. Freeing the bolt.

e. **Bolt Group.** After freeing the bolt, slide it to the rear and out of receiver (Figure 2-11). Place the bolt down on its right side (with the extractor arm up), so that the extractor will not fall from the bolt.

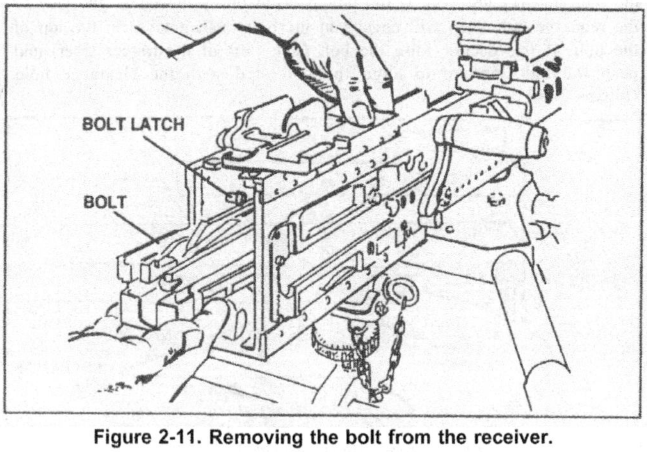

Figure 2-11. Removing the bolt from the receiver.

f. **Barrel Buffer Body Group and Barrel Extension Group.** Insert the drift of a combination tool, or other pointed instrument, through the hole in the lower rear corner of the right sideplate. Push in on the barrel buffer body lock. At the same time, place one hand in the receiver and push the barrel extension group and barrel buffer group to the rear (Figure 2-12). Remove the barrel buffer group and barrel extension group from the receiver. Separate the two groups by pushing forward on the tips of the accelerator (Figure 2-13).

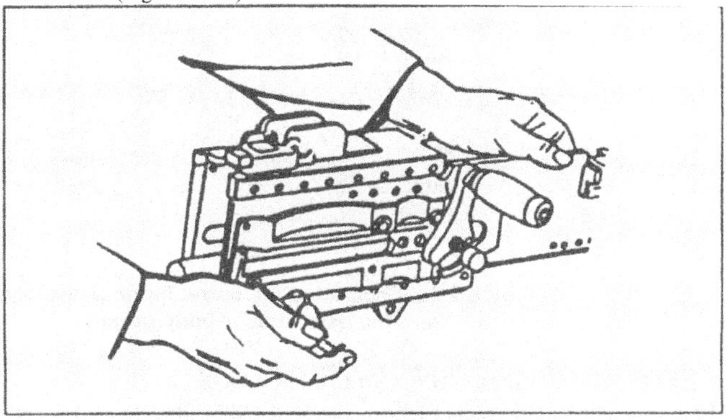

Figure 2-12. Removing barrel buffer group and barrel extension group.

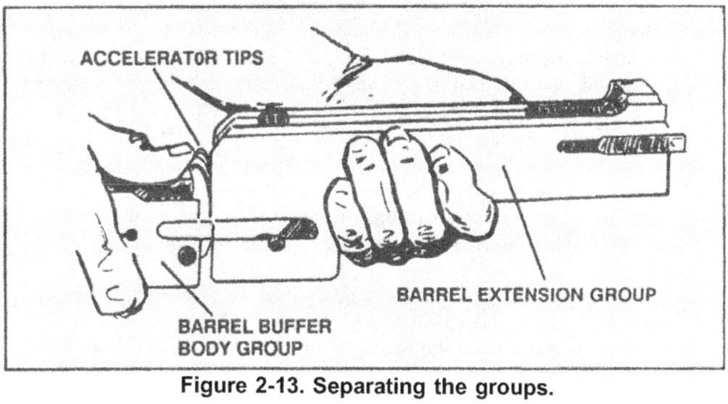

Figure 2-13. Separating the groups.

g. **Barrel Buffer Assembly.** Pull the barrel buffer assembly from the rear of the barrel buffer body group. The barrel buffer assembly will not be disassembled (Figure 2-14). This completes general disassembly.

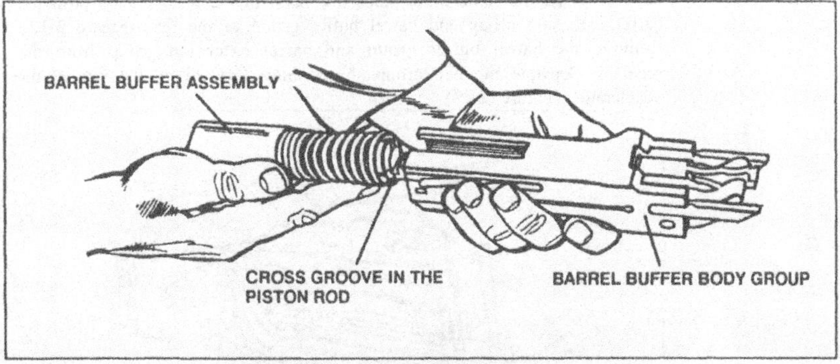

Figure 2-14. Separating the barrel buffer assembly from the barrel buffer body group.

2-3. CLEANING, INSPECTION, AND LUBRICATION

To ensure proper care of the MG, it is necessary to have a system of maintenance or an SOP for the frequency of cleaning. Each gun should be cleaned as soon after firing as possible and each time it is exposed to field conditions. Under combat conditions, the gun should be cleaned and oiled daily. Under extreme climatic and combat conditions, it maybe necessary to clean and lubricate more frequently. Under ideal conditions, where the gun is not used, and is stored in a clean, dry place, it may only be necessary to inspect, clean, and lubricate every 5 days. The gun should be disassembled, cleaned, and oiled in a clean, dry location. If possible, keep the gun covered with a gun cover, canvas, tarpaulin, or poncho when not in use.

a. **Routine Care and Cleaning.** Before firing (when the situation permits), take the following steps to ensure efficient functioning of the machine gun:

- Disassemble the gun into its major groups or assemblies.
- Clean the bore and chamber, and lightly oil them.
- Clean all metal parts thoroughly with CLP. See paragraph 2-3f for lubrication procedures.

b. **Care and Cleaning Under Unusual Conditions.** Extreme cold, hot, dry, and tropical climates affect the gun and its functioning. Care should be taken under these climatic conditions to ensure that the gun is cleaned daily with the prescribed lubricants and protected from the elements by some sort of cover if possible. Further information on care and cleaning of the gun under unusual climatic conditions can be found in TM 9-1005-213-10.

c. **Care and Cleaning of M3 Mount and Accessories.** The mount and accessories, such as the ammunition chest and spare parts, should also be kept clean and lubricated. Painted surfaces should be spot painted when necessary. Moving surfaces should be inspected and oiled with the prescribed lubricant. All external surfaces of the mount should be kept clean and lightly oiled. Be particularly careful that the pintle bushing is clean and lightly oiled, and that the pintle lock release cam is well-lubricated and free from grit. The sleeve lock indexing levers and telescopic legs should be clean and lubricated enough for ease in use. The mount should be cleaned and oiled with the same regularity and in the same manner as the gun.

d. **Maintenance and Inspection.** Units must establish guidelines and conduct regular maintenance and inspection to keep the machine gun and its mounts in operational conditions.

(1) *Gun maintenance.* The importance of a thorough knowledge of care, cleaning, and maintenance of the machine gun cannot be overemphasized because these actions determine whether or not the gun will function properly when needed. The bore and chamber must be properly maintained to preserve accuracy. Because of the close fit of working surfaces and the high speed at which the gun operates, the receiver and moving parts must be kept clean, correctly lubricated, and free from burrs, rust, dirt, or grease to ensure proper, efficient functioning.

(2) *Mount maintenance.* The care, cleaning, lubrication, and adjustment of the mounts used with the gun are no less important. The functioning of the gun and mount together determine overall effectiveness. All accessories and equipment used with the gun and mount, including ammunition, must also be properly maintained.

(3) *Inspection.* When inspected, the machine gun should be completely disassembled. Inspecting personnel should look for dirt, cracks, burrs, and rust.

e. **Inspection Checklist.** Table 2-1 is an inspection checklist to be used as a guide for crewmembers or inspecting personnel to ensure that the gun and equipment are properly maintained.

UNIT	INSPECTION
1. GUN.	
a. Barrel.	Inspect the bore and chamber for rust. See that they are clean and lightly oiled.
b. Moving parts.	See that they are clean and lightly oiled. Operate the retracting slide handle and bolt latch release several times to see that the parts function without excessive friction.
c. Headspace and timing.	Check with the gauges to ensure that headspace and timing are correct.
d. Rear sight and windage knob.	Ensure that the sight is in good condition, clean, free of grease or dirt, and lightly oiled. Elevation should be set at 1,000, windage zero, and the sight should be down.
2. MOUNT (M3, MK64, M36, or M4).	See that it is clean, lightly lubricated, and that all clamps are securely tightened. It should function properly and be complete.
3. SPARE PARTS AND TOOLS.	Inspect to see that they are clean and lightly oiled. See that spare parts kits are complete and in good condition. Replacement parts should be requisitioned and newly drawn parts examined.
4. T&E.	See that it is clean, lightly lubricated, and that both hand wheels work properly.
5. AMMUNITION.	See that ammunition is properly stored, and that boxes and ammunition are in good condition and not oiled.

Table 2-1. Inspection checklist.

f. **Lubrications.** Use cleaner, lubricant, preservative to clean the machine gun. As its name implies, it cleans, lubricates, and preserves all in one application.

(1) After cleaning the gun with CLP, wipe it dry and reapply a thin coating. Allow this thin coat to dry on the parts for a short time before reassembly. CLP deposits a thin coating on the metal which minimizes carbon buildup and prevents foreign material from sticking. It is this coating that provides the frictionless operation of the weapon parts, not liquid oil deposited on them. A gun treated with CLP will operate better and remain clean longer than one treated with any other cleaning material. Use of CLP will reduce maintenance costs and extend the life of the weapon.

(2) Rifle bore cleaner is a cleaning solvent which can be used to clean powder residue, carbon, and dirt from weapons. RBC does not preserve or lubricate a weapon. If you clean a weapon with RBC, dry the weapon and lubricate it with lubricating oil, semifluid (LSA); lubricating oil, special purpose (PL-S); or lubricating oil, general purpose (PL-M). The use of these oils will cause sand or grit to stick to the weapon. RBC and oil should be used only when CLP is not available.

2-4. MAINTENANCE PROCEDURES

There are certain actions that must be taken before, during, and after firing to properly maintain the gun. See Table 2-2 for a preoperation checklist.

PART	BEFORE	DURING FIRING OR TEMPORARY CESSATION	AFTER
Bore	Make sure it is clear and clean.		Clean and oil lightly.
Moving parts	Oil lightly and test for worn or broken parts. They should function without excessive friction.	Lubricate working parts. Observe the functioning of the gun to anticipate failures.	Inspect, clean, and oil lightly.
Headspace and timing	Check adjustment by gauge and correct if necessary.	Watch for bulged cases to prevent a ruptured (separated) case. If a separated case occurs, remove it and readjust headspace.	Check adjustment.
Rear sight and windage knob	See that the sight is clean and functions properly. Set sight at 1,000, windage zero.	Keep properly set.	Clean and oil; set at 1,000, windage zero.
Spare parts and tools	Clean and oil spare parts and tools. Check kits for completeness. Examine newly drawn parts.	Keep available.	Clean and oil. Check and replace damaged or missing parts.
Ammunition	Have an adequate supply; clean, correctly loaded, and in good condition. DO NOT OIL.	Keep correctly aligned with the feedway; check resupply. Protect from sun, moisture, and dirt. Watch for link stoppage.	Clean, store carefully, and replenish supply.

Table 2-2. Operational checklist.

2-5. MAINTENANCE UNDER NBC CONDITIONS

If contamination is anticipated, apply oil to all outer metal surfaces of the weapon. DO NOT OIL AMMUNITION. Keep the weapon covered as much as possible. If the weapon is contaminated, decontaminate it as prescribed by FM 3-5 and then clean and lubricate it.

2-6. GENERAL ASSEMBLY

To assemble the gun, replace the groups and assemblies in reverse order of their removal in disassembly.

a. **Barrel Buffer Assembly and Barrel Buffer Body Group.** Replace the barrel buffer assembly in the barrel buffer body group, with the key on the spring guide to the right. This key must fit in its slot in the right side of the barrel buffer body. Turn the barrel buffer tube until the screwdriver slot (in the rear of the tube) is vertical and the arrow is pointing to the right. The stud on the tube lock will now engage the serrations in the barrel buffer tube to keep the tube from turning. Push the barrel buffer assembly fully forward (Figures 2- 15 and 2-16).

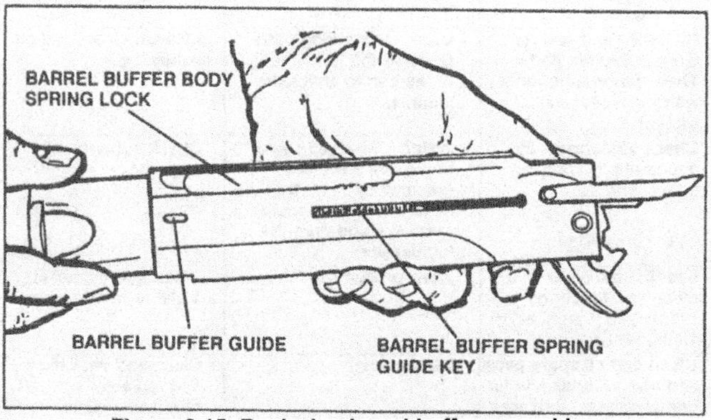

Figure 2-15. Replacing barrel buffer assembly.

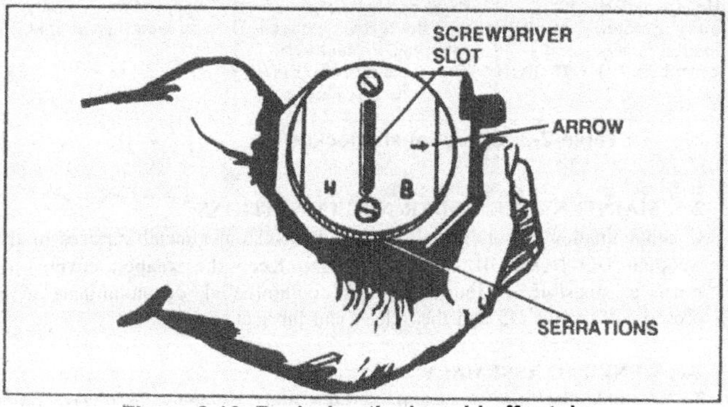

Figure 2-16. Replacing the barrel buffer tube.

b. **Barrel Buffer Group and Barrel Extension Group.** To join the two groups together, hold the barrel buffer group in the right hand, with the index finger supporting the accelerator. Join the notch on the shank of the barrel extension group with the cross-groove in the pistol rod of the barrel buffer assembly. At the same time, align the breech lock depressors with their guideways in the sides of the barrel extension, ensuring that the tips of the accelerator are against the rear end of the barrel extension (claws against the shank) (Figure 2-17). Push the groups together. As the accelerator rotates to the rear, press down on its tips to ensure positive locking of groups. Place the groups in the receiver, and push them forward until the barrel buffer body spring lock snaps into position. When the parts are properly locked in place, the barrel buffer tube should protrude about 1 1/8 inches from the rear of the barrel buffer body group.

c. **Bolt.** Place the bolt in the receiver, with the top of the cocking lever forward and the extractor down. The barrel extension, barrel buffer, and bolt groups may be assembled and returned to the receiver together (Figure 2-18).

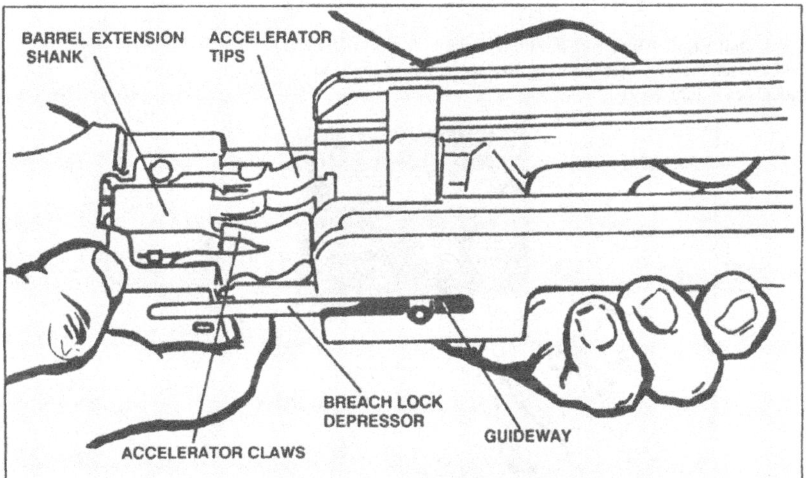

Figure 2-17. Joining the barrel extension group and barrel buffer group.

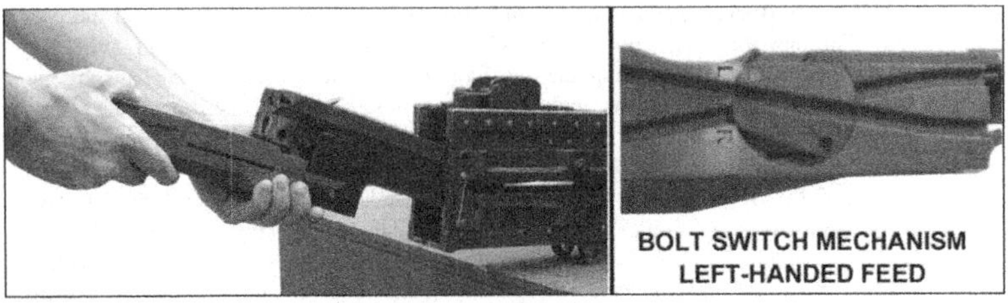

*** Figure 2-18. Returning the barrel extension, barrel buffer, and bolt groups together.**

*** CAUTION**

Before inserting the bolt group, ensure bolt switch mechanism is positioned on **L** (left-hand) feed for the weapon.

* d. **Bolt Stud**. Align the stud hole in the bolt with the clearance hole and replace the bolt stud, ensuring that the collar of the stud is inside the sideplate (Figure 2-19).

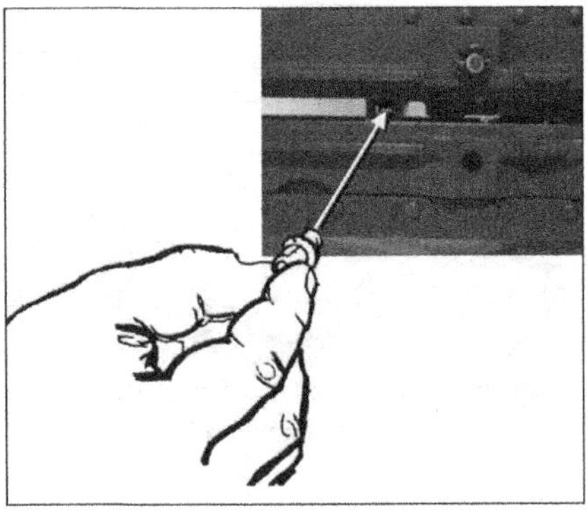

*** Figure 2-19. Replacing the bolt stud.**

* e. **Drive Spring Assembly**. Press up on the bolt latch and push the bolt all the way forward by pushing on the bolt stud only. Place the end of the drive spring rod in its hole in the rear of the bolt, and push forward on the drive spring assembly and the barrel buffer tube. Press in and to the right on the head of the drive spring rod and place the retaining pin in its seat in the right sideplate (Figure 2-20).

* Figure 2-20. Inserting drive spring and drive spring rod (with drive spring rod inside drive spring).

* NOTE: At this time, the barrel buffer tube should be completely inside the receiver. If not, the barrel buffer body spring is not properly seated.

* f. **Backplate Group**. Hold the backplate with the latch down and the trigger up; place the backplate guides in their guideways. Hold out on the latch lock and tap the backplate into position until the latch snaps into place (Figure 2-21). Release the latch lock and pull up on the backplate group to ensure it is firmly seated.

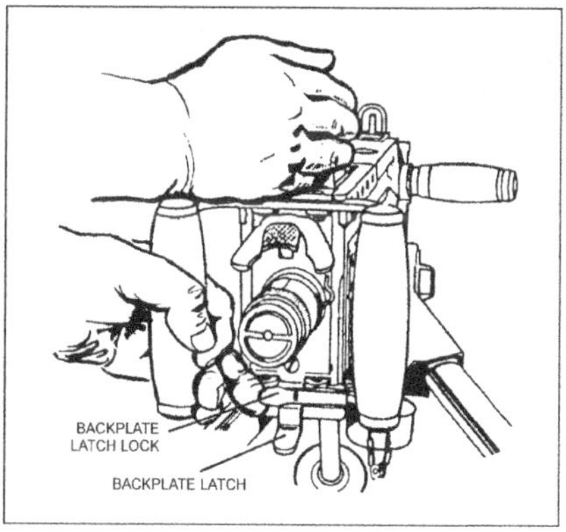

* Figure 2-21. Replacing the backplate group.

*** CAUTION**

Do not use the driving rod to drive the bolt forward from the rear position. This may damage the drive spring group and cause a stoppage.

* g. **Barrel.** Pull the retracting slide handle to the rear until the lug on the barrel locking spring is visible through the 3/8-inch hole in the right sideplate. Place the smallest loop of a caliber .50 link, or suitable spacer, between the trunnion block and the barrel extension. Screw the barrel all the way into the barrel extension; then unscrew the barrel two notches. Remove the link and close the cover. This completes general assembly.

2-7. FUNCTION CHECK

A function check must be performed as soon as the weapon is assembled to ensure that the weapon has been assembled correctly. The following procedures should be taken to check the function of the weapon.

a. Place the weapon in the single-shot mode.

b. Open the cover and lock the bolt to the rear (bolt should stay to rear while in the single-shot mode).

c. Hold the retractor handles, press the bolt latch release, and ride the bolt forward.

d. Press down on the trigger; weapon should fire. (Check T-slot to ensure that firing pin does protrude.)

e. Place the weapon in the automatic-fire mode

f. Pull the retractor slide handle to the rear and hold. (Bolt should not lock to rear.)

g. Release the pressure on the slide handles and ride the bolt forward

h. Make sure firing pin does not protrude

i. Press trigger; weapon should fire

j. Make sure firing pin does protrude.

NOTE: Before firing a newly assembled weapon, first set the headspace and timing. Chapter 3 describes these procedures.

* CHAPTER 3

OPERATION AND FUNCTIONING

This chapter explains the operation of the MG. It discusses the loading, unloading, and clearing procedures, and the cycle of functioning of the weapon. When training the cycle of functioning using dummy ammunition, it is imperative that all safety procedures be followed.

* 3-1. OPERATION

The overall operation of the MG includes how to load, unload, and clear the weapon. During the weapon's operation, it is mandatory that all ammunition be free of dirt and corrosion, that the ammunition be properly linked, and that the double-linked end be at the top of the ammunition can.

* 3-2. LOADING PROCEDURES

Trainers must ensure that the weapon functions correctly and that proper headspace and timing have been set before loading. When loading in either mode, the ammunition is fed into the MG in the same manner (Figure 3-1). Ensure that the bolt is forward and the cover is closed. Insert the double-loop end of the ammunition belt into the feedway until the first round is engaged by the belt-holding pawl. Figure 3-2 shows the correct position of the bolt latch in the single-shot or automatic mode.

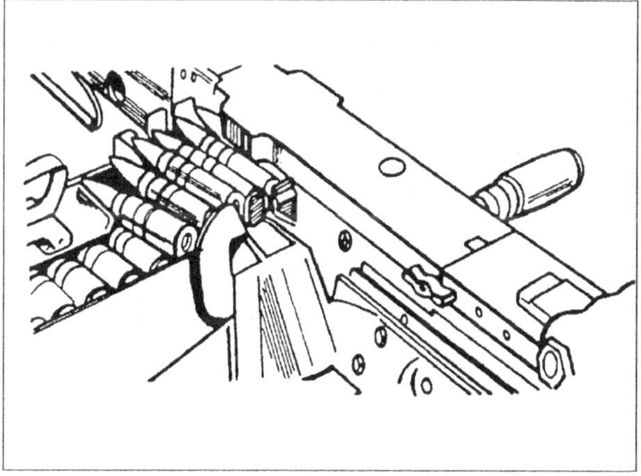

Figure 3-1. Inserting ammunition.

a. **Single-Shot Mode**. When engaging targets at ranges greater than 1,100 meters, using the single-shot mode (firing one round at a time) allows the gunner to deliver well-aimed fire on the target. To load in the single-shot mode--

(1) Keep the bolt-latch release unlocked in the up position and release it manually for each round.

(2) Jerk the retracting slide handle to the rear and lock it in position. Return the retracting slide handle to the forward position and then release the bolt by pressing the bolt latch release. The gun is now half-loaded.

(3) To complete loading, jerk the retracting slide handle to the rear and lock it in position. Return the retracting slide handle to the forward position. Press the bolt latch release. When the bolt goes forward for the second time, the gun is loaded.

b. **Automatic Mode**. To load in the automatic mode--

(1) Lock the bolt-latch release down with the bolt- latch release lock.

(2) Jerk the retracting slide handle to the rear and release it. The gun is now half-loaded.

(3) To complete loading, jerk the retracting slide handle to the rear a second time and release it. When the bolt goes forward for the second time, the gun is loaded.

Figure 3-2. Firing modes.

* 3-3. UNLOADING PROCEDURES

To unload the MG, the gunner must first ensure that the weapon is in the single-shot mode. The cover is then lifted and the assistant gunner removes the ammunition belt from the feedway. The bolt is then locked to the rear. If a round is chambered, it will release, unfired, when the bolt locks to the rear. Once the bolt is locked to the rear, the chamber and

T-slot are examined to ensure that they are not holding rounds. In darkness, this must be done by feeling the areas. After the examination has been done (during training), a wooden block is inserted in the receiver between the bolt and the rear of the barrel, extending above and below the receiver about one inch. Then a cleaning rod is inserted in the muzzle end of the barrel and pushed through the bore until it can be seen in the receiver. Remove the rod, the gun is now clear (Figure 3-3).

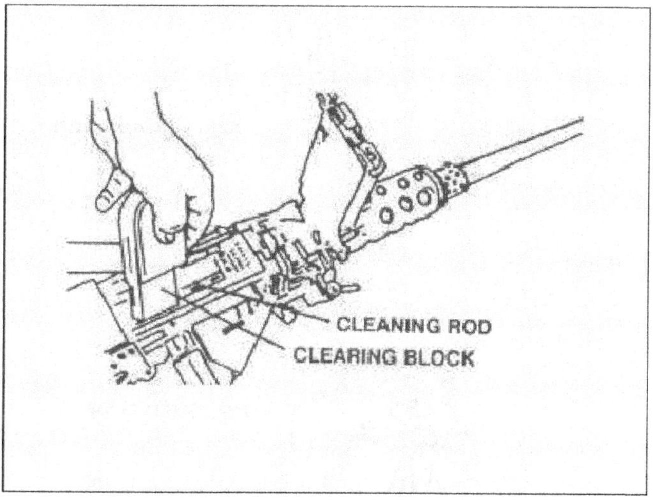

Figure 3-3. The clearing block.

* **3-4. CYCLE OF FUNCTIONING**

The cycle of functioning is broken down into basic steps: feeding, chambering, locking, firing, unlocking, extracting, ejecting, and cocking. Some of these steps may occur at the same time.

a. **Feeding**. Feeding is the act of placing a cartridge in the receiver, approximately in back of the barrel, ready for chambering. When the bolt is fully forward and the top is closed, the ammunition belt is held in the feedway by the belt-holding pawl (Figure 3-4).

(1) As the bolt is moved to the rear, the belted ammunition is moved over and then held in a stationary position by the belt-holding pawl. At the same time, the belt-feed pawl rides up and over the link, holding the first round in place. When the bolt is all the way to the rear, the belt-feed slide moves out far enough to allow the belt-feed pawl spring to force the pawl up between the first and second rounds (Figure 3-5).

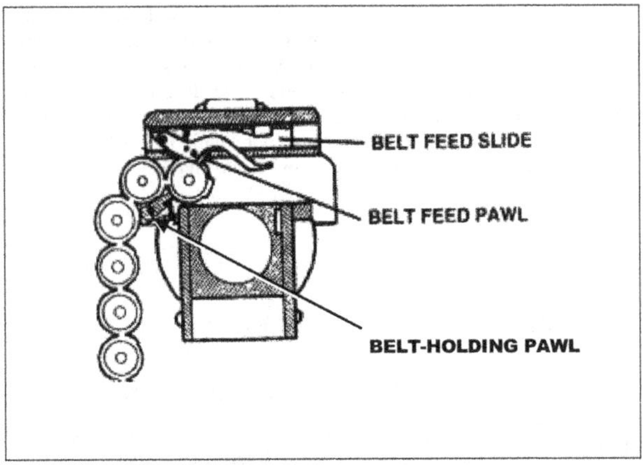

Figure 3-4. Feeding--step 1.

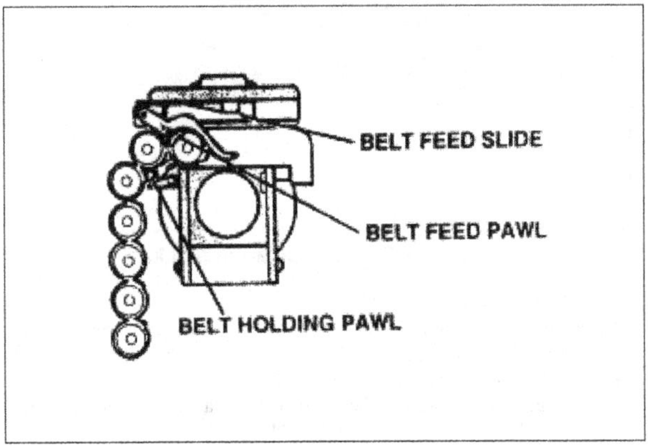

Figure 3-5. Feeding--step 2.

(2) As the bolt moves forward, the belt-feed slide is mo
the receiver, pulling with it the next linked cartridge. Wh
reaches the fully forward position, the belt-holding pawl will
place behind the second linked cartridge (Figure 3-7), holding it
The extractor will then grasp the rim of the first cartridge, prepa
release it from the belt on the next rearward motion (Figure 3-8).

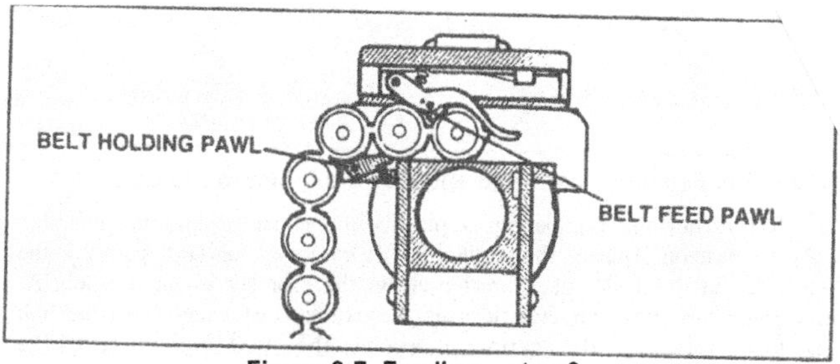

Figure 3-7. Feeding – step 3.

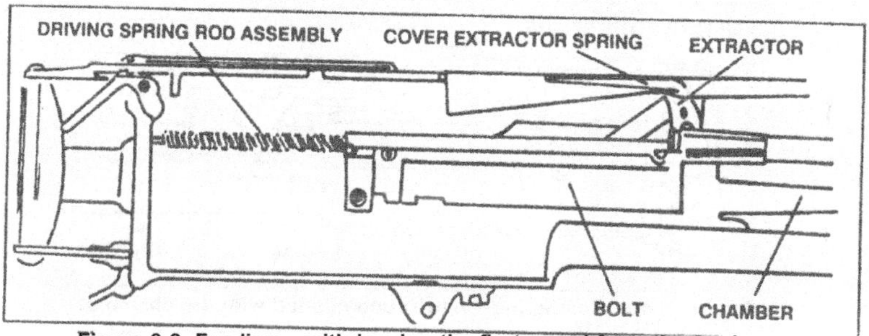

Figure 3-8. Feeding – withdrawing the first round from the feedway.

(3) As the bolt then moves to the rear, the extractor will pull the cartridge with it, releasing it from the belt. As it moves to the rear, the extractor is forced down by the extractor cam, causing the cartridge to be moved into the T-slot in the bolt face, preparing the cartridge to be chambered (Figure 3-9, page 3-6). It is connected under the extractor switch on the side of the receiver until it is repositioned by the forward movement of the bolt, and pressure of the cover extractor spring forces it over the next round.

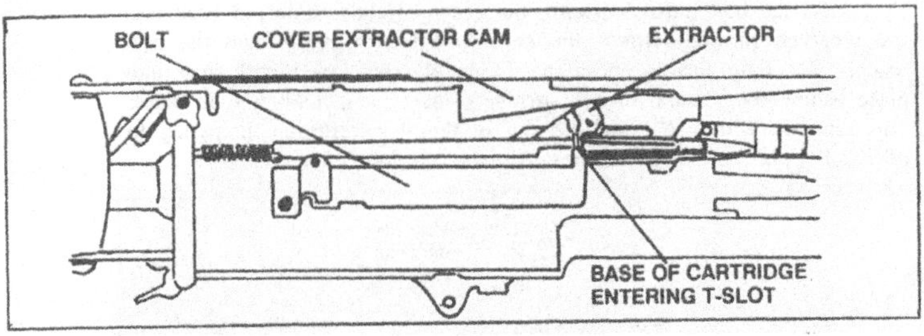

Figure 3-9. Feeding - cartridge entering the T-slot in the bolt.

b. **Cambering.** Cambering is placing the cartridge into the chamber of the weapon. During this cycle, the bolt moves forward, carrying the cartridge in the T-slot in a direct route to the chamber of the weapon. At the same time, the extractor rides up the extractor cam and when the bolt is fully forward, the extractor grasps the next linked cartridge (Figure 3-10).

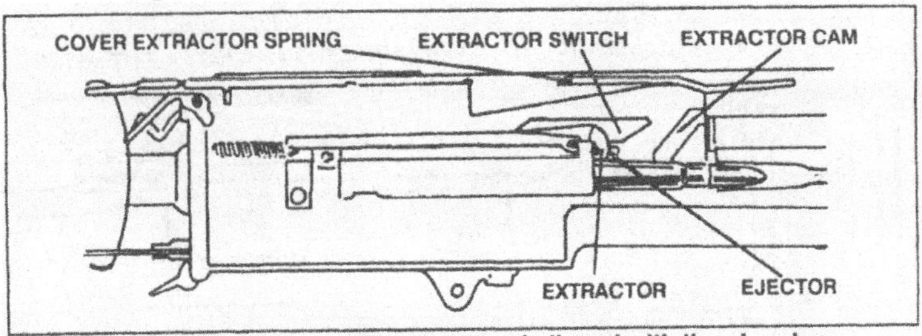

Figure 3-10. Cambering – new round aligned with the chamber.

c. **Locking.** The bolt is locked to the barrel and barrel extension.

(1) Initially, the bolt is forced forward in counter-recoil by the energy stored in the driving spring assembly and the compressed buffer disks. At the start of counter-recoil, the barrel buffer body tube lock keeps the accelerator tips from bounding up too soon and catching in the breech lock recess in the bolt. After the bolt travels forward about 5 inches, the lower rear projection of the bolt strikes the tips of the accelerator, turning the accelerator forward. This unlocks the barrel extension from the barrel buffer body group and releases the barrel buffer spring. The barrel buffer spring expands, forcing the piston rod forward.

(2) Since the cross groove in the piston rod engages the notch on the barrel extension shank, the barrel extension and barrel are also forced forward by the action of the barrel buffer spring. Some of the forward motion of the bolt is transmitted to the barrel extension through the accelerator. As the accelerator rotates forward, the front of the accelerator speeds up the barrel extension; at the same time, the accelerator tips slow down the bolt.

(3) Locking begins 1 1/8 inches before the recoiling groups (bolt, barrel extension, and barrel) are fully forward. The breech lock in the barrel extension rides up the breech lock cam in the bottom of the receiver into the breech lock recess in the bottom of the bolt, locking the recoiling groups together. The recoiling groups are completely locked together three-fourths of an inch before the groups are fully forward (Figure 3-11).

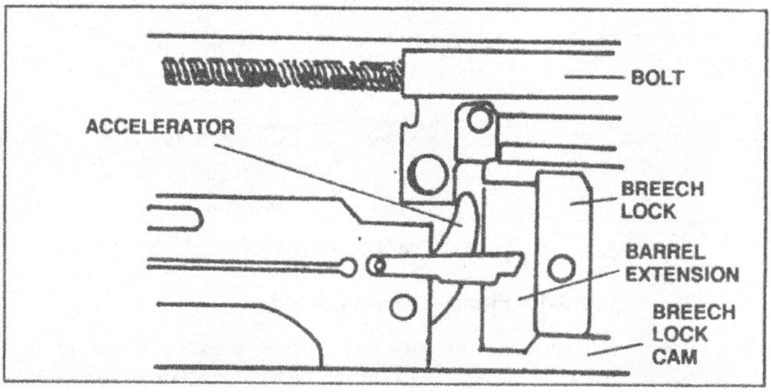

Figure 3-11. Locking – recoiling groups locked together.

d. **Firing.** The firing pin is released, igniting the primer of the cartridge.

(1) As the trigger impressed down, it pivots on the trigger pin, so that the trigger cam on the inside of the backplate engages and raises the rear end of the trigger lever. This in turn pivots on the trigger lever pin assembly, causing the front end of the trigger lever to press down on the top of the sear stud. The sear is forced down until the hooked notch of the firing pin extension is disengaged from the sear notch. The firing pin and firing pin extension are driven forward by the firing pin spring; the striker of the firing pin hits the primer of the cartridge, firing the round (Figures 3-12 and 3-13, page 3-8).

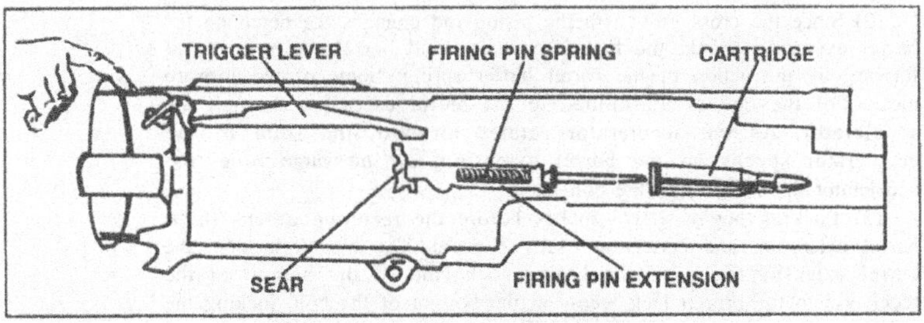

Figure 3-12. Firing – ready to fire.

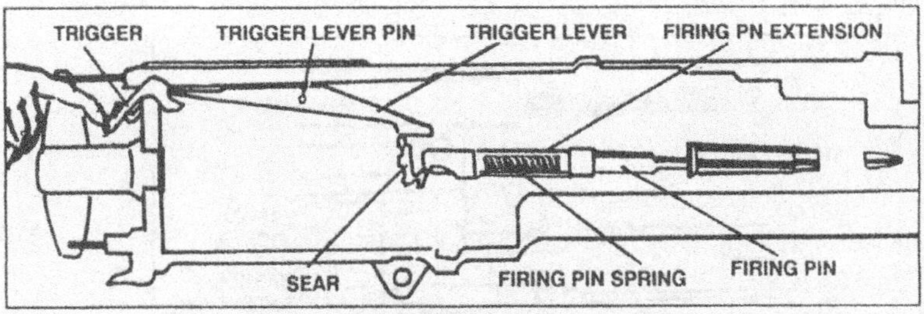

Figure 3-13. Firing – round ignited.

(2) For automatic firing, the bolt-latch release must be locked or held depressed, so that the bolt latch will not engage the notches in top of the bolt, holding the bolt to the rear as in single-shot firing. The trigger is pressed and held down. Each time the bolt travels forward in counter-recoil, the trigger lever depresses the sear, releasing the firing pin extension assembly and the firing pin. This automatically fires the next round when the forward movement of the recoiling groups is nearly completed. The gun should fire about one-sixteenth of an inch before the recoiling groups are fully forward. Only the first round should be fired with the parts fully forward. The gun fires automatically as long as the trigger and bolt latch are held down and ammunition is fed into the gun.

e. **Unlocking.** The bolt is unlocked from the barrel and barrel extension.

(1) At the instant of firing, the bolt is locked to the barrel extension and against the rear end of the barrel by the breech lock, which is on top

of the breech lock cam and in the breech lock recess in the bottom of the bolt. When the cartridge explodes, the bullet travels out of the barrel; the force of recoil drives the recoiling groups rearward. During the first three-fourths of an inch, the recoiling groups are locked together. As this movement takes place, the breech lock is moved off the breech lock cam stop, allowing the breech lock depressors (acting on the breech lock pin) to force the breech lock down, out of its recess from the bottom of the bolt. At the end of the first three-fourths of an inch of recoil, the bolt is unlocked, free to move to the rear independent of the barrel and barrel extension.

(2) As the recoiling groups move to the rear, the barrel extension causes the tips of the accelerator to rotate rearward. The accelerator tips strike the lower rear projection of the bolt, accelerating the movement of the bolt to the rear. The barrel and barrel extension continue to travel to the rear an additional three-eighths of an inch, or an approximate total distance of 1 1/8 inches until they are stopped by the barrel buffer assembly (Figure 3-14).

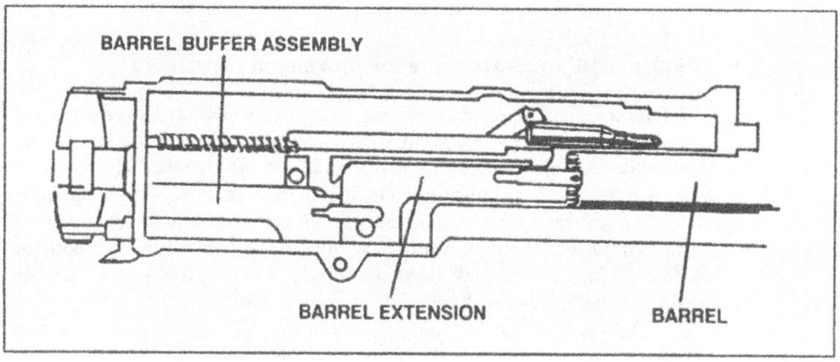

Figure 3-14. Unlocking – barrel and barrel extension stopped by the barrel buffer assembly.

(3) During the recoil of 1 1/8 inches, the barrel buffer spring is compressed by the barrel extension shank, since the notch on the shank is engaged in the cross groove in the piston rod head. The spring is locked in the compressed position by the claws of the accelerator, which engage the shoulders of the barrel extension shank. After its initial travel of three-fourths of an inch, the bolt travels an additional 6 3/8 inches to the rear, after it is unlocked from the barrel and barrel extension, for a total of

7 1/8 inches. During this movement, the driving springs are compressed. The rearward movement of the bolt is stopped as the bolt strikes the buffer plate. Part of the recoil energy of the bolt is stored by the driving spring rod assembly, and part is absorbed by the buffer disks in the backplate (Figure 3-15).

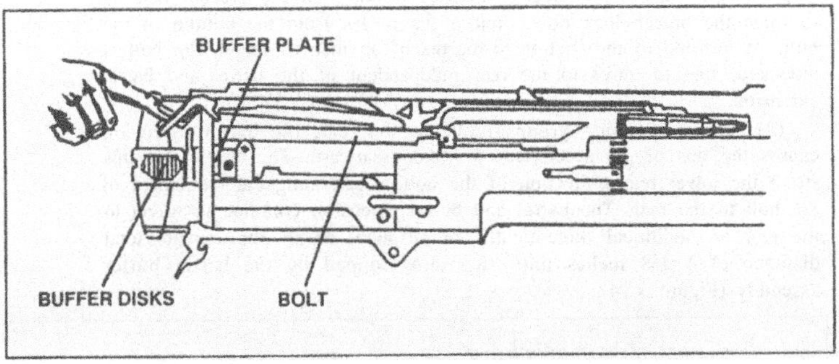

Figure 3-15. Unlocking – recoil movement completed.

f. **Extracting.** The empty cartridge case is pulled from the chamber.

(1) The empty case, held by the T-slot, has been expanded by the force of the explosion; therefore, it fits snugly in the chamber. If the case is withdrawn from the chamber too rapidly, it may be torn. To prevent this, and to ensure slow initial extraction of the case, the top forward edge of the breech lock and the forward edge of the lock recess in the bolt are beveled. As the breech lock is unlocked, the initial movement of the bolt away from the barrel and barrel extension is gradual.

(2) The slope of the locking faces facilitates locking and unlocking and prevents sticking. The leverage of the accelerator tips on the bolt speeds extraction after it is started by kicking the bolt to the rear to extract the empty case from the chamber.

g. **Ejecting.** The empty cartridge case is expelled from the receiver.

(1) As the bolt starts its forward movement (counter-recoil), the extractor lug rides below the extractor switch, forcing the extractor assembly farther down until the round is in the center of the T-slot of the bolt.

(2) The round, still gripped by the extractor, ejects the empty case from the T-slot. The last empty case of an ammunition belt is pushed out by the ejector.

h. **Cocking.** The firing pin is withdrawn into the cocked position.

(1) When the recoiling groups are fully forward, the top of the cocking lever rests on the rear half of the V-slot in the top plate bracket. As the bolt moves to the rear, the top of the cocking lever is forced forward. The lower end pivots to the rear on the cocking lever pin. The rounded nose of the cocking lever, which fits through the slot in the firing pin extension, forces the extension to the rear, compressing the firing pin spring against the sear stop pin (accelerator stop). As the firing pin extension is pressed to the rear, the hooked notch of the extension rides over the sear notch, forcing the sear down. The sear spring forces the sear back up after the hooked notch of the firing pin extension has entered the sear notch.

(2) The pressure of the sear and firing pin springs holds the two notches locked together. There is a slight overtravel of the firing pin extension in its movement to the rear to ensure proper engagement with the sear. As the bolt starts forward, the overtravel is taken up and completed when the cocking lever enters the V-slot of the top plate bracket, and is caromed toward the rear; pressure on the cocking lever is relieved as the bolt starts forward.

3-5. LEFT-HAND FEED

By repositioning some of the components, the MG is capable of alternate feed. Ammunition can be fed into the weapon from the right or left side of the receiver; however the Army uses only left-hand feed. (See Table 3-1).

PART	POSITION
Belt-feed lever, shoulder headless pin, and spring.	Upper (rear) hole; lug of feed lever is on left side of cover.
Belt-feed slide.	Feed pawl is on left side of cover.
Belt-feed pawl arm.	Arm toward latch end of cover, pointing right.
Cover latch shaft lever	Left side of cover.
Cartridge stops and link stripper.	Right-hand rear cartridge stop assembly and front cartridge stop on right side of feedway.
Retracting slide with handle.	Right sideplate.
Belt-holding pawl.	Left side of feedway.
Bolt switch.	Cam grove in line with "L" on bolt.

Table 3-1. Position of parts for left-hand feed.

DANGER

Clear the gun of ammunition before setting headspace.

3-6. HEADSPACE AND TIMING

Headspace is the distance between the face of the bolt and the base of the cartridge case, fully seated in the chamber. Timing is the adjustment of the gun so that firing takes place when the recoiling parts are in the correct position for firing. Because the cartridge is held by the T-slot of the bolt, headspace with the MG is measured as the distance between the rear of the barrel and the face of the bolt. This occurs when the recoiling parts are forward and there is positive contact between the breech lock recess in the bolt and the lock in the barrel extensions. Periodic calibration checks should be made of the gauge by direct support personnel at least annually.

WARNING

Firing a weapon that has improperly set headspace and timing could result in damage to the machine gun, or injury to the gunner. Damage may also occur in the trunnion block, base of the barrel, or face of the bolt. This warning applies whether the gun is firing service ammunition or M1E1 blanks. (The weapon has improper early timing when two rounds are fired – and firing stops.)

a. **Gauges.** The headspace and timing gauge consists of a headspace gauge and two timing gauges (Figure 3-16). These gauges provide an accurate means of checking the adjustment of headspace and timing.

NOTE: The headspace and timing gauge should be kept with the gun at all times.

Figure 3-16. Headspace and timing gauge.

b. **Headspace.** Check and set headspace before firing, after assembling the gun, and after replacing the barrel or receiver group. Use the following procedures to set headspace.

(1) Raise the cover all the way up. Grasp the retracting slide handle (Figure 3-17). Using the retracting slide handle, retract the bolt until the barrel-locking-spring lug is centered in the 3/8-inch hole on the right side of the receiver (Figure 3-18).

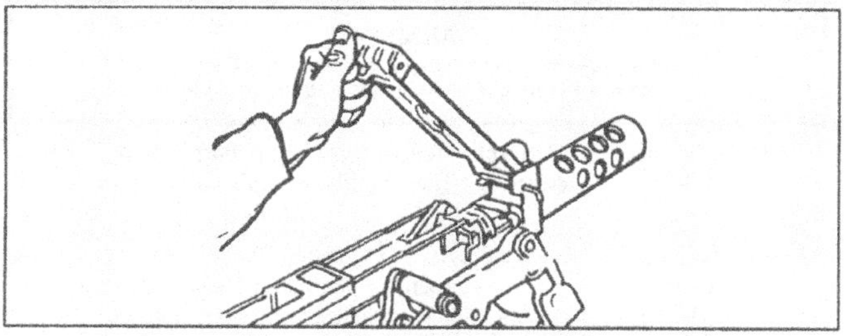

Figure 3-17. Raising the cover.

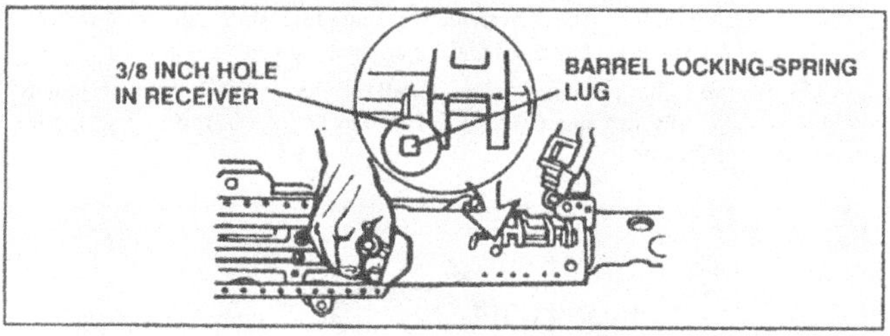

Figure 3-18. Retracting the bolt.

(2) Hold the bolt in this position and screw the barrel fully into the barrel extension (Figure 3-19).

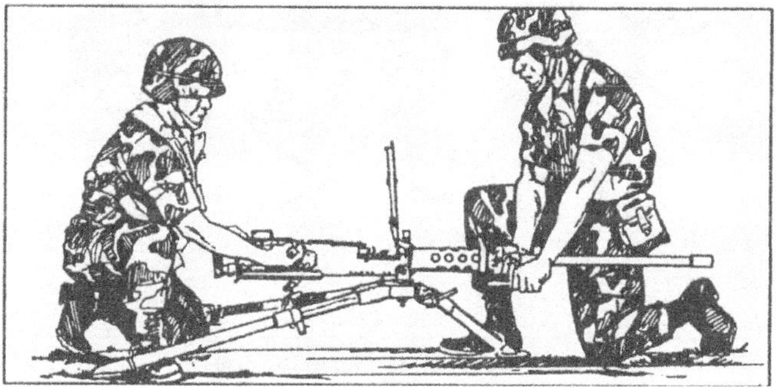

Figure 3-19. Screwing in the barrel.

WARNING
When resetting the headspace and timing of a gun that has been fired, use an asbestos mitt to avoid burns.

(3) With the bolt still retracted, unscrew the barrel two notches (clicks). Release the retracting slide handle and allow the bolt to go forward.

NOTE: At this point, check the barrel for rotation. Attempt to turn the barrel in either direction. The barrel should not turn. If the barrel does turn, stop here and check barrel notches and the barrel-locking spring for damage.

(4) Pull the bolt to the rear with the retracting slide handle and hold. This cocks the weapon. Otherwise, the headspace gauge will not fit.

(5) Holding the retracting slide handle, release the bolt, and slowly return the bolt fully forward (Figure 3-20). Do not press the trigger or let the bolt slam forward.

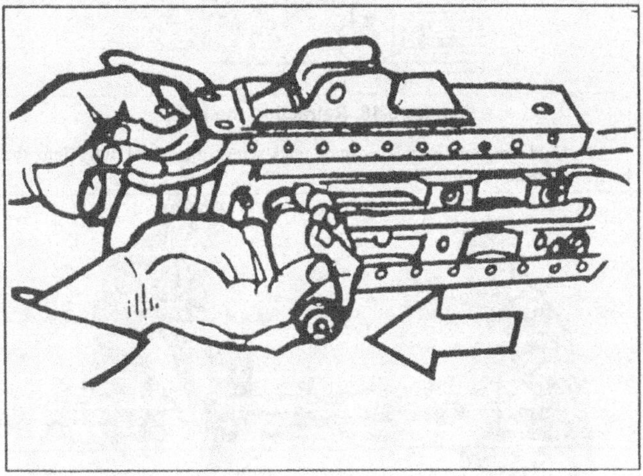

Figure 3-20. Releasing the bolt.

(6) Retract the retracting slide handle and separate the barrel extension from the trunnion block by a 1/16-inch gap (Figure 3-21).

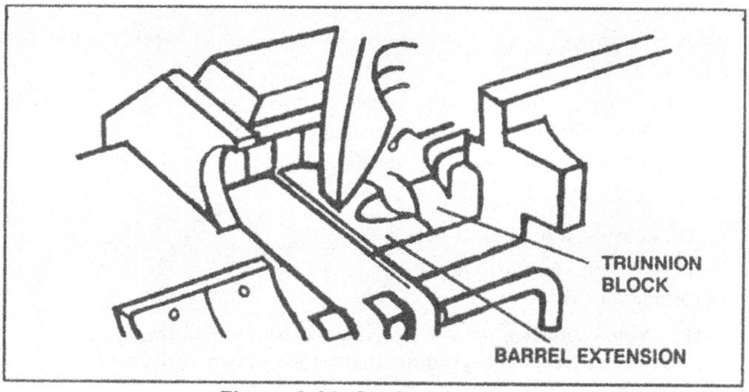

Figure 3-21. Setting the gap.

(7) Raise the extractor out of the way to clear the top of the T-slot and try both ends of the go/no-go gauge as shown in Figure 3-22.

NOTE: Insert the go end of the gauge between the face of the bolt and the end of the barrel all the way up to the ring.

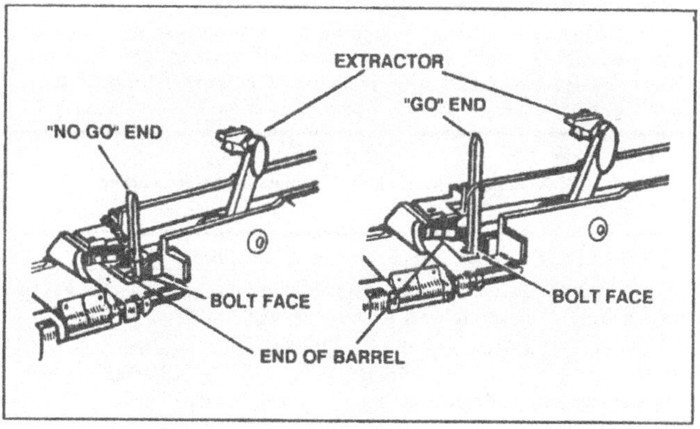

Figure 3-22. Using the gauge.

(8) If the go end of the gauge enters the T-slot freely to the center ring of the gauge, and the no-go end will not enter, headspace is correct. Remove gauge. Headspace setting is now complete. If the go end of the gauge will not enter the T-slot freely, headspace is too tight. Continue as follows.

(9) Retract the bolt so you can see the barrel-locking lug in the center of the 3/8th-inch alignment hole on the right side of the receiver.

(10) Unscrew the barrel one notch (click).

(11) Return the bolt fully forward.

(12) Recheck headspace (step 9).

(13) Repeat steps 10 through 13 until the go gauge fits but the no-go gauge does not fit.

NOTE: You should not have to unscrew the barrel more than five notches (clicks) beyond the first setting of two clicks. If this condition does occur, turn in the machine gun to your unit armorer for inspection.

(14) If the no-go end of the gauge enters the T-slot, headspace is too loose. Adjust it using the same procedures as above, screwing the barrel into the barrel extension rather than out.

(15) Repeat steps 10 through 13, one click at a time, until the no-go gauge does not fit but the go gauge does fit.

c. **Timing.** Timing is the adjustment of the weapon so that firing takes place when the recoiling parts are between .020 and .116 inch out of battery to prevent contact between the front end of the barrel extension and the trunnion block. Use the following procedures to set timing.

WARNING

Make sure the gun is clear of ammunition before starting.

(1) Check headspace first as previously described.

(2) Pull the bolt to the rear with the retracting slide handle and then ease bolt fully forward to cock the machine gun.

(3) Grasp the retracting slide handle and retract the bolt just enough (1/16 inch) to insert the no-fire gauge between the barrel extension and the trunnion block. Release the retracting slide handle (Figure 3-23).

(4) Depress the trigger. Gun should not fire.

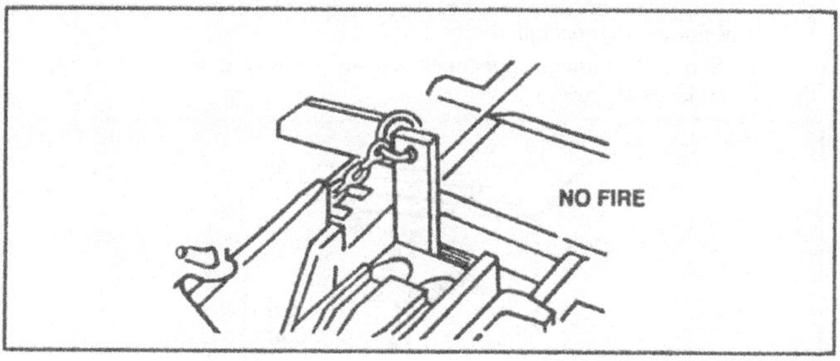

Figure 3-23. Inserting the no-fire gauge.

NOTE: Insert timing gauge with bevel against barrel notches.

NOTE: If the gun does not fire, go to step 5. If the gun does fire, you have early timing. Go on to steps 7 through 14.

(5) Grasp the retracting slide handle and retract the bolt just enough to remove the no-fire gauge and insert the fire gauge in the same place (Figure 3-24). Release the retracting slide handle.

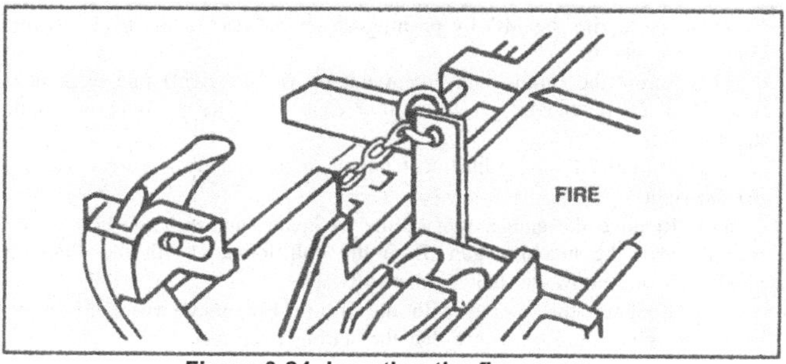

Figure 3-24. Inserting the fire gauge.

(6) Depress the trigger. Gun should fire. If it does, timing adjustment is now complete.

NOTE: If the gun does not fire, you have late timing. Go to steps 7 through 14.

(7) Remove the gauge, cock the gun, and return the bolt forward

(8) Insert the fire gauge.

(9) Remove the backplate.

(10) Screw the timing adjustment nut all the way down until it touches the trigger lever (Figure 3-25).

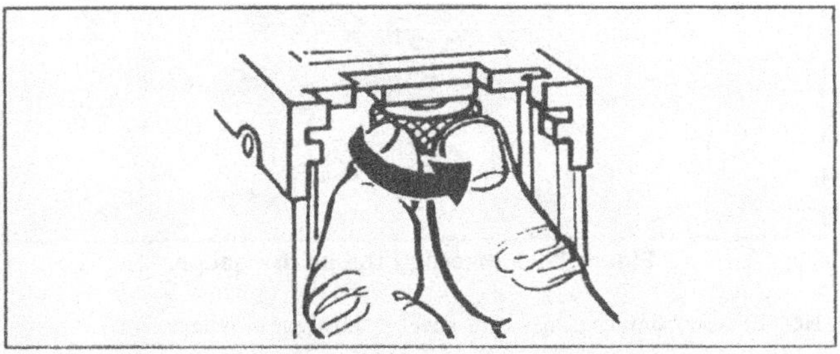

Figure 3-25. Adjusting the timing nut.

> **WARNING**
> **Never cock the gun with the backplate off.**

(11) Try to fire the MG by pushing up on the rear of the trigger lever. Gun should not fire.

(12) Screw the timing adjustment nut up (to the right) one click at a time. Push up on the trigger lever after each click. Keep doing this until the gun fires.

(13) Turn the timing adjustment nut no more than two more clicks up (to the right).

(14) Remove the gauge, replace the backplate, and pull the bolt to the rear to cock the machine gun. Ease the bolt forward with the charging handle. Do not allow the bolt to slam forward.

(15) Recheck the timing with the fire/no-fire gauge twice after the backplate is installed to ensure that the setting is correct.

d. **Field Expedient Methods.** When a go/no-go gauge is not available, you can still set the headspace and timing using field expedient methods. However, this method should be used only in combat.

(1) *To set headspace:*

(a) Raise the cover and retract the bolt in the normal manner until the barrel-locking-spring lug is centered in the 3/8-inch hole on the right side of the receiver.

(b) Hold the bolt in this position and screw the barrel fully into the barrel extension; then unscrew the barrel two clicks or notches.

(2) *To set timing:*

(a) Use a dog tag or a dime as a fire gauge.

(b) Use a nickel and a dime or four dog tags as a no-fire gauge.

(c) Set the timing using the normal procedure.

(3) *To check for correct settings:*

(a) Attempt to fire the weapon. If it fires sluggishly, clear the weapon then unscrew the barrel one more notch.

(b) Recheck the rate of fire. Repeat the procedures in paragraph (1); however, do not exceed two more clicks.

(c) Do not unscrew the barrel more than one notch between test firings.

CHAPTER 4

PERFORMANCE PROBLEMS AND DESTRUCTION

This chapter discusses the malfunctions, stoppages, immediate actions, remedial actions, and destruction procedures of the MG. All corrective actions must comply with safety procedures.

4-1. MALFUNCTIONS

A malfunction is any failure of the gun to function satisfactorily. Examples of malfunctions are:

a. **Failure to Function Freely.** Sluggish operation is usually due to human failure to eliminate excessive friction caused by dirt, lack of proper lubrication, burred parts, incorrect headspace adjustment, or incorrect timing.

b. **Uncontrolled Automatic Fire.** Uncontrolled automatic fire (runaway gun) is when fire continues even when the trigger or trigger control mechanism is released. If the cause is present before the gun is fired, the gun will start to fire when the recoiling groups move into battery the second time. If the defect occurs during firing, the gun will continue firing when the trigger control mechanism is released. A runaway gun may be caused –

- By a bent trigger lever, forward end of the trigger lever sprung downward.
- By burred beveled contacting surfaces of the trigger lever and sear.
- By a jammed or broken sideplate trigger.

To stop the uncontrolled automatic fire:

(1) Keep the gun laid on target.

(2) Twist the belt, causing the gun to jam.

(3) Caution, do not unlatch the cover.

(4) Wait 5 minutes to guard against cook off.

(5) Clear weapon, replace broken, worn, or burred parts. Check the sideplate trigger and trigger control mechanism, when applicable.

4-2. STOPPAGES

A stoppage is any interruption in the cycle of operation caused by the faulty action of the gun or ammunition. Stoppages are classified as follows:

a. **Failure to Feed.** Prevents the round from being properly positioned in the receiver group.

b. **Failure to Chamber.** Prevents the complete chambering of the round.

c. **Failure to Lock.** Prevents the breech lock from correctly entering its recess in the bolt.

d. **Failure to Fire.** Prevents the ignition of the round.

e. **Failure to Unlock.** Prevents the breech lock from moving out of its recess in the bolt.

f. **Failure to Extract.** Prevents the extraction of the expended cartridge from the chamber.

g. **Failure to Eject.** Prevents the ejection of the expended cartridge from the receiver.

h. **Failure to Cock.** Prevents the firing pin extension from being engaged with the sear.

Table 4-1 lists the causes of most stoppages.

NATURE OF STOPPAGE	USUAL CAUSES	OTHER CAUSES
Failure to feed.	Defective ammunition belt. Defective feed mechanism. Defective extractor.	Improperly loaded belt. Short round
Failure to chamber.	Broken part or obstruction in T-slot or chamber. Separated (ruptured) case.	Thick or thin rim, bulged round, protruding primer.
Failure to lock	Incorrect headspace.	Broken parts. Battered breach lock. Battered breach lock cam. Faulty breech lock cam adjustment.
Failure to fire.	Defective parts in firing mechanism. Defective ammunition. Incorrect timing.	
Failure to unlock.	Broken parts in receiver.	Worn or faulty breech lock cam, or faulty adjustment.
Failure to extract.	Dirty chamber.	Defective cartridge case.
Failure to eject.	Defective ejector.	Burred T-slot.
Failure to cock.	Broken sear. Worn sear notch. Weak sear spring. Worn hooked notch on firing pin extension.	Broken cocking lever.

Table 4-1. Stoppages and their causes.

4-3. IMMEDIATE ACTION

The first thing to do when a stoppage occurs is to apply immediate action, which is the prompt action taken by the firer to reduce a stoppage.

Immediate action is performed by the gunner; however, every crewmember must be trained to apply immediate action. The following procedures will assist in reducing most stoppages without analyzing their cause in detail.

WARNING

Failure to follow procedures may result in damage to weapon and injury to personnel.

a. If gun fails to fire, take the following action:

(1) Wait 5 seconds; a hangfire maybe causing the misfire. In the next 5 seconds, pull the bolt to the rear (check for ejection and feeding of belt), release it, re-lay on the target, and attempt to fire. If the bolt-latch release and trigger are depressed at the same time, the bolt goes forward and the weapon should fire automatically.

(2) If the gun again fails to fire, wait 5 seconds, pull the bolt to the rear (engage with bolt latch if applicable), and return the retracting slide handle to its forward position. Open the cover and remove the belted ammunition. Inspect the gun to determine the cause of stoppage.

b. A hangfire or cookoff can cause injury to personnel or damage to the weapon. To avoid these, the gunner must take the following precautions:

(1) Always keep the round locked in the chamber the first 5 seconds after a misfire occurs. This prevents an explosion outside of the gun in event of a hangfire.

(2) If the barrel is hot, the round must be extracted within the next 5 seconds to prevent a cookoff. When more than 150 rounds have been fired in a 2-minute period, the barrel is hot enough to produce a cookoff.

(3) If the barrel is hot and the round cannot be extracted within the 10 seconds, it must remain locked in the chamber for at least 5 minutes, to guard against a cookoff.

(4) Keep the gun cover closed during the waiting periods.

4-4. REMEDIAL ACTION

When immediate action does not correct the malfunction, the quickest way to resume firing is to replace the defective part.

a. **Removal of a Cartridge from the T-Slot.** If the cartridge does not fall out, hold the bolt to the rear, and with the extractor raised, use a screwdriver to push the cartridge out the bottom of the receiver.

b. **Removal of a Ruptured Cartridge.** A ruptured (separated) cartridge case may be removed with a cleaning rod or ruptured cartridge extractor. When using the ruptured cartridge extractor, raise the cover and pull the bolt to the rear. Place the extractor in the T-slot of the bolt in the same manners as that of a cartridge, so that it is held in line with the bore by the ejector of the extractor assembly of the gun. With the extractor aligned with the bore and held firmly in the T-slot, let the bolt go forward into the ruptured case, and the shoulders will spring out in front of the case. Pull the bolt to the rear and remove the ruptured case and extractor (Figures 4-1 and 4-2).

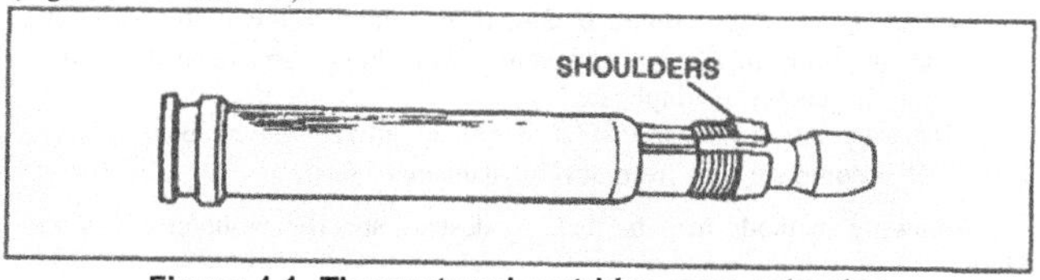

Figure 4-1. The ruptured cartridge case extractor.

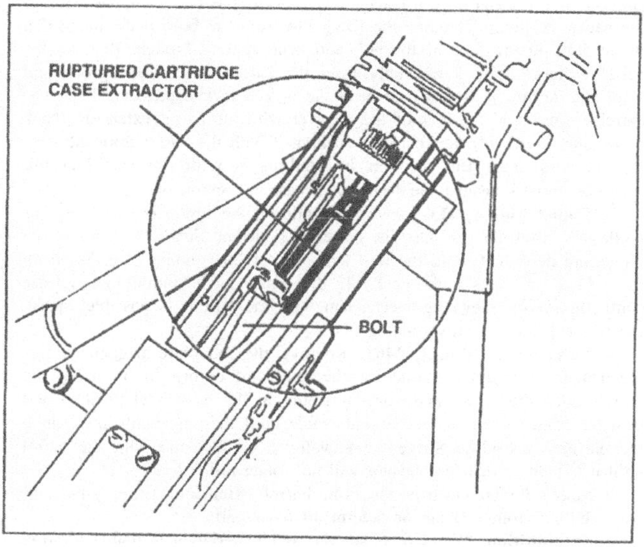

Figure 4-2. Ruptured cartridge case extractor aligned with the chamber.

4-5. DESTRUCTION PROCEDURES

The decision to destroy the gun to prevent its capture and use by the enemy is a command decision. It will be ordered and carried out only on authority delegated by the major unit commander.

- Destroy the machine gun and mount only when they are subject to capture or abandonment. Destruction must be as complete as circumstances permit.
- Lacking time for complete destruction, destroy the parts essential to operation of the gun, beginning with those parts most difficult for the enemy to duplicate.
- Destroy the same parts of each gun to prevent the reconstruction of a complete gun from several damaged guns.

The following methods may be used to destroy specific components of the MG.

a. **Gun.** Field strip the gun. Use the barrel as a sledge. Raise the cover and smash the cover forward and down toward the barrel support. Smash the backplate group. Remove the firing pin from the bolt; place the striker in the hole in the face of the bolt and bend it until broken. Remove the barrel buffer tube lock assembly from the barrel buffer body group and bend and deform it. Smash and bend the breech lock depressors. Place the barrel extension in the rear of the receiver with the barrel extension shank protruding; knock off the shank by striking it with the barrel from the side. Deform and crack the receiver by striking it with the barrel at the sideplate corners nearest the feedway. Smash the extractor.

b. **Tripod Mount, M3.** Leave the pintle on the tripod by removing the pintle bolt from the gun. Use the barrel as a sledge. Strike the sides of the pintle and deform it. Fold the trail legs and turn the mount over. Stand on the folded trail legs and knock off the pintle latch (pintle lock release cam). Smash the elevating mechanism with the barrel. If possible, smash the rear legs to prevent unfolding.

c. **Antiaircraft Mount, M63.** Remove the sideplate trigger control mechanism from its container or the gun, and deform it by using the barrel. Lock the cradle and yoke assembly in the horizontal position and beat the trigger frame assembly and cradle until they are bent down along the elevator assembly. Strike the elevator from the side with the barrel until it is bent so that the elevator will not rotate in the base.

d. **Spare Parts.** Destroy the bolt, barrel extension, firing pins, and barrel buffer groups. Break or deform all other parts.

e. **Ammunition.** When time permits and material is available, destroy ammunition by burning. Unpack all ammunition from boxes or cartons, stack the ammunition in a heap and, using flammable material available, ignite and take cover immediately.

f. **Burning.** To destroy the gun by burning, place a thermite grenade in the receiver on the bolt (with the cover resting on the grenade) and fire the grenade. (This method may require the use of more than one grenade.) Remove the backplate group, place a thermite grenade in the rear of the receiver and fire the grenade.

g. **Disposal.** Bury in suitable holes, or dump parts into streams, mud, snow, sumps, latrines, or scatter the parts over a wide area.

CHAPTER 5

MARKSMANSHIP TRAINING

The procedures, exercises, and techniques for implementing a proper marksmanship program are based on the concept that all machine gunners must understand common firing principles, be proficient marksmen, and be confident in applying these skills in combat. This depends on their understanding of the machine gun and their application of marksmanship fundamentals. Proficiency is accomplished through practice that is supervised by qualified trainers and through objective performance amassments by the unit leaders. The structure of this chapter is in four sections: planning, fundamentals, basic marksmanship, and advanced gunnery. All advanced exercises are conducted under conditions that are as much like tactical conditions as possible.
(See Appendix B for training aids and devices.)

Section I. PLANNING

The planning of MG training is no different from other marksmanship training. Guidelines are provided to assist the trainers in understanding, preparing, and ensuring that all training is conducted to standard. This section addresses the objectives, the responsibilities of the commanders, and the phases of training. It also introduces the training devices that assist in training, and designates when remedial and sustainment training should be conducted.

5-1. OBJECTIVES

The objectives of the MG training program are to guide the trainers, leaders, and gunners through a sequence of training to standard that produces a gunner who is able to maintain the gun and effectively engage targets in combat.

5-2. COMMANDER'S RESPONSIBILITIES

The responsibilities of the commander are to ensure the instructors and their assistants are thoroughly trained and rehearsed in the planning, knowledge, and presentation of all MG training. He will ensure that safety is emphasized and never overlooked during training. Serviceable weapons are a must for good training and commanders must make sure they are used properly. He also ensures sufficient time is scheduled.

5-3. PHASES OF TRAINING

The initial training strategy is divided into three phases of marksmanship instruction — the fundamental phase, the basic marksmanship phase, and the advanced gunner phase. Gunner safety is continuously stressed during all phases of training.

a. In the fundamental phase, the gunner learns the necessary common skills, such as dry-fire exercises, preparation of positions, manipulation of the T&E, range determination, and sighting and aiming. The gunner must master these skills before he is allowed to progress to the basic marksmanship phase.

b. During the basic marksmanship phase, the gunner learns the basics of loading, zeroing, laying, and engaging of single and multiple targets from the tripod and vehicle mode.

c. The advanced gunnery phase trains the gunner in engaging moving targets, night firing, NBC firing, mounted firing, and firing using fire commands. The gunner will be placed under the stress and strain of simulated combat conditions.

5-4. SUSTAINMENT TRAINING

Once individuals and units have trained to a required level of proficiency, leaders must structure collective and individual training plans to conduct critical task training at the frequency necessary for the sustainment training strategy. Mission training plans and individual training evaluation programs help achieve and sustain collective/individual proficiency. Sustainment training prevents skill degradation. Army units must be prepared to accomplish their wartime missions at any time – they cannot rely on infrequent peaking to the appropriate level.

5-5. REMEDIAL TRAINING

Remedial training will be conducted in any of the phases of training where the gunner does not meet the standard. The trainer must be instantly aware of any gunner that seems to be having trouble. Once the problem has been identified, the gunner should be retrained as soon as possible so that he will maintain the same level of proficiency as the other gunners.

Section II. FUNDAMENTALS

The fundamentals are necessary basic skills that a gunner must be trained in before he can be expected to effectively engage targets. Personnel conducting marksmanship training must fully understand the fundamentals and be well rehearsed in applying them. The basics in MG training are assuming a proper firing position, sighting, aiming, determining range, and manipulating the T&E mechanism.

5-6. FIRING POSITIONS

Before a gunner can hit targets, he must learn to get behind the weapon in a position that allows him to be comfortable, affords him protection, and enhances mission accomplishment.

a. The tripod firing positions are prone, sitting, and standing. They are assumed in the following manner.

(1) The prone position is used when firing from the tripod that is set in a low position. It is assumed by lying on the ground directly behind the gun. The gunner then spreads his legs a comfortable distance apart with his toes turned outward. His left elbow rests on the ground, and his left hand grasps the elevating handwheel of the T&E. His right hand lightly grasps the right spade grip with his right thumb in a position to press the trigger. The position of his body can then be adjusted to position his firing eye in alignment with the sights of the weapon (Figure 5-1).

Figure 5-1. Prone position with tripod mount.

(2) The sitting position can be used when the tripod is set in a high or low position. The gunner sits directly behind the gun between the legs of the tripod. He may extend his legs under the tripod or cross them, depending on his physique. The gunner then places both elbows on the inside of his thighs to get the best support. He grasps the elevating

handwheel of the T&E with the left hand, and lightly grasps the right spade grip with his right hand. He must ensure that the right thumb is in position to press the trigger (Figure 5-2).

Figure 5-2. Sitting position with tripod mount.

(3) The standing position is used when the gunner is firing from a fighting position. This position is assumed by standing directly behind the gun with the feet spread a comfortable distance apart. The gunner grasps the elevating handwheel of the T&E with the left hand. He lightly grasps the right spade grip with the right hand, ensuring that the right thumb is in a position to press the trigger. Adjustment of the body is allowed in order to align the firing eye with the sights on the weapon (Figure 5-3).

Figure 5-3. Standing position with tripod mount.

b. The vehicular firing position for the MG is standing. It is assumed by constructing a solid platform to stand on, using sandbags or ammunition boxes; or, in the case of the M113 APC, using the commander's seat. The gunner must then ensure that his platform is high enough to place the spade grips of the gun about chest high. He grasps the spade grips with both hands and places both thumbs in a position to press the trigger. The gunner holds the gun tightly to his chest for stabilization; his elbows should be locked tightly to his sides. He sights over the weapon and adjusts his position by flexing his knees and leaning forward to absorb any recoil (Figure 5-4).

Figure 5-4. Standing position using cupola mount.

c. The antiaircraft firing position uses a standing position when firing from the M63 mount. To assume the position, the gunner stands with his feet spread comfortably apart with his shoulders squarely behind the gun (Figure 5-5, page 5-6). When the gunner is engaging aerial targets, he grasps the upper extension handles with both hands. When engaging low-level aircraft or ground targets, he grasps the lower extension handles with both hands.

NOTE: The kneeling position may be used; it has the advantage of presenting a lower profile of the gunner and also aligns the gunner's eye closer to the axis of the barrel.

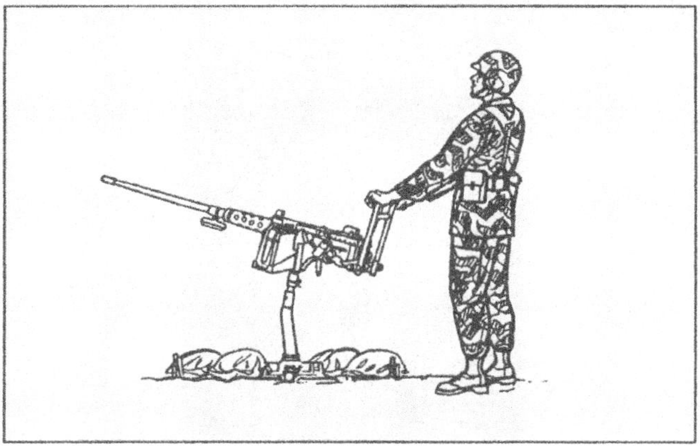

Figure 5-5. Antiaircraft firing position.

5-7. DRY-FIRE TRAINING

Dry-fire training is designed to teach the gunner the essentials of MG gunnery, including safety. Dry-fire training also includes sighting, aiming, sight setting, laying, manipulating the gun, manipulating the T&E mechanism, and determining the range. Thorough, carefully supervised training of these essentials is necessary to conserve time and ammunition during live fire. Practical exercises should be used to determine gunners' proficiency. Mastery of these skills is a must before the gunner is allowed to move on to the next phase of training. Practice is a must to achieve mastery.

a. **Sighting and Aiming.** Sighting is the ability of the gunner to use correct sight alignment and correct sight picture to engage targets.

(1) The first step in proper sighting is finding a natural, comfortable spot where the gunner is able to see the front sight blade through the rear peep sight. It is important the gunner understands that the spot he chooses to sight from must be constant throughout his firing.

(2) The second step in sighting is to move the weapon until the top center of the front sight blade is exactly in the center of the rear sight peep hole. The gunner can achieve this by drawing imaginary lines that bisect in the center of the rear peephole and then placing the top of the front ight blade center of them. This is correct sight alignment (Figure 5-6).

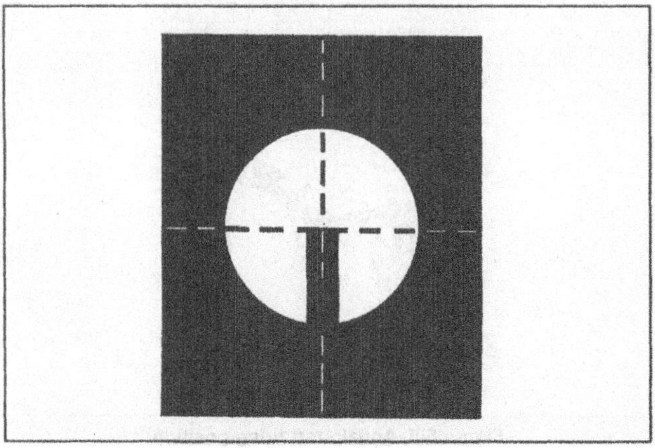

Figure 5-6. Correct sight alignment.

(3) The third step is to establish correct sight picture. To perform this task, the gunner adjusts the weapon until the top center of the front sight blade is bottom center of the intended target. Correct sight picture is a combination of sight alignment and placement of point of aim (Figure 5-7).

Figure 5-7. Correct sight picture.

b. **Range Setting and Laying.** Range setting and laying the gun are important elements in marksmanship training. It is this training that prepares the gunner to accurately and rapidly place fire on his target in combat. To properly set ranges, the gunner must be trained in rear sight operation.

(1) Setting ranges on the rear sight is a simple but important task. The gunner just has to rotate the elevating screw knob in a clockwise manner to move the peep sight up or counterclockwise to move it down. The range scale on the left is graduated in mils and the scale on the right in yards. The gunner must align the hairline index of the peep sight with the scale index line at the desired range as quickly and accurately as possible.

(2) Laying is placing the barrel of the weapon on a direct line with the target, using the sights. This must also be done as quickly and accurately as possible.

(3) The range setting and laying exercises are designed to require the gunner to practice and the instructor to evaluate both correct sight alignments and correct placements of aiming points. The exercise starts with the gunner in the sitting firing position with rear sight down. The coach will announce a range and general aiming point. The gunner will then repeat the range and direction of target. The coach will then announce, "Begin." The gunner will then raise and set his sights, and align the weapon on the aim point. When the gunner completes this task, he will announce "Up." The coach will then get behind the weapon and check the range setting and aim point and critique the gunner on his findings. This exercise will be continued until speed and accuracy is obtained.

c. **Traversing and Elevating Mechanism.** Manipulation of the T&E mechanism (Figure 5-8) is another key factor in effectively engaging targets. The gunner is taught how to instinctively manipulate the T&E mechanism to shift from one target to another. The gunners are trained to use the traversing handwheel, the traversing slide lock, and the elevating handwheel.

(1) The traversing mechanism consists of a traversing handwheel, locking nut, scale, and yoke. The T&E mechanism is attached to the traversing bar of the M3 bipod.

(a) The traversing bar is graduated in 5-mil increments and fits between the trail legs of the M3 tripod. The traversing slide and screw assembly are clamped in place on the traversing bar by the traversing slide lock lever. When the traversing slide is locked to the traversing bar, the traversing handwheel should be centered. The traversing slide is properly mounted when the lock lever is to the rear and the traversing handwheel is positioned to the left.

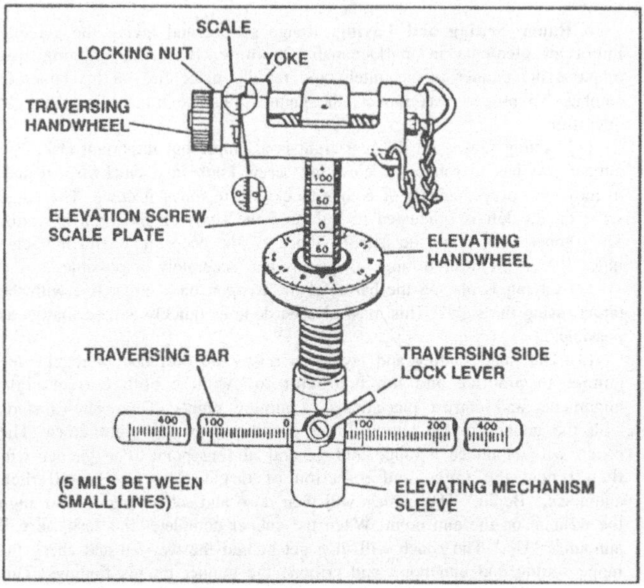

Figure 5-8. Traversing and elevating mechanism.

(b) To make changes in direction, loosen the traversing slide lock lever and move the slide along the traversing bar. This permits traverse of 400 mils left or right of the zero index in the center of the traversing bar. Readings on the traversing bar are taken from the left side of the traversing slide. For changes of 50 mils or less in deflection, turn the traversing handwheel. This allows a traverse of 50 mils left or right of center. One click in the traversing handwheel signifies 1 mil change in direction.

(2) The elevating mechanism consists of an upper and lower elevating screw, which is connected to the gun by inserting the quick release pin assembly through the holes in the upper elevating screw yoke and the rear mounting lugs of the receiver. A scale, graduated in mils, is fitted to the upper screw to indicate elevation. This scale is marked to show (-) minus 250 mils in depression and (+) plus 100 mils in elevation from the zero setting.

(3) The elevating handwheel is graduated in 1 mil increments from 0 to 50 mils and is fastened to the elevating screw by a screw lock. This synchronizes the handwheel graduations with those on the upper elevating screw. A spring-actuated index device produces a clicking sound when the handwheel is turned. Each click equals 1 mil change in elevation. The handwheel is turned clockwise to depress the barrel and counterclockwise to elevate.

(4) Direction and elevation readings constitute the data necessary to engage preselected target areas during limited visibility. These readings are measured by and recorded from the traversing bar and the T&E mechanism. To obtain accurate readings, the T&E must be first zeroed with all measurements recorded in mils.

(a) To zero the traversing handwheel, the gunner must first hold the T&E so that the traversing handwheel is on his left as he looks at it. He then turns the handwheel toward himself until it stops, loosens the locking nut slightly, and aligns the zero on the scale with the zero on the elevating screw yoke. Once the zeros are aligned, he tightens the locking nut. He must then turn the handwheel two complete turns away from the body and stop. The scale should again be on the zero. If this procedure is done at night, the gunner will turn 50 clicks away from him.

(b) To zero the elevating handwheel, the gunner must first turn the handwheel up or down until the handwheel is level with the line directly under the zero on the elevating screw plate scale, and the elevating handwheel indicator is pointing to the zero on the top of the handwheel. The elevating mechanism sleeve is then rotated up until it is stopped by the handwheel. The gunner then rotates the sleeve down until it stops, making sure he counted each complete rotation. He then divides the number of rotations by two, rotates the sleeve back up that number, and stops. The T&E mechanism is now ready to be attached to the tripod.

(c) To obtain and record direction readings, the gunner sets the sight on the proper range to hit the target, loosens the traversing slide lock lever, and slides the T&E mechanism along the traversing bar until the weapon is sighted on the aiming point of the target. The T&E mechanism is then locked down by tightening the traversing slide lock lever. All readings are taken from the left side of the sleeve mechanism. If the left side of the sleeve is not exactly on one of the 5-mil tick marks, the gunner must slide the sleeve to the next smaller tick mark to align it exactly. The traversing handwheel is then used to move the weapon back on point of aim. The direction is now ready to be recorded. The reading is taken from the number on the traversing bar and the direction from the direction of the barrel of the weapon. If the sleeve mechanism is on the right side of the zero on the traversing bar, then the reading is left; if it is on the left side of the zero, then it is a right reading. The width of a target may also be measured and recorded by first moving the traversing handwheel until the sights are aligned with the right or left side of the target. The clicks required to do this measure the width.

NOTE: Before repositioning the weapon for another target, the gunner must realign the handwheel.

(d) To obtain an elevation reading, the gunner must first ensure that the sights are aligned and at the desired aim point of the target. The elevation reading is made up of two portions, a major reading and a minor reading. The major reading is taken from the elevating screw plate scale. The scale is graduated in 50-mil increments and ranges from a minus (-) 250 mils to a plus (+) 100 mils with a zero between them. There is an index line below each number and a plus or minus sign above each number, with the exception of the zero. The zero does not have a plus or minus sign. To obtain the elevation reading, the gunner should lower his head until his eyes are level with the elevating handwheel. The major reading is the first number with a plus or minus sign, with its index line just visible above the elevating handwheel. The minor reading is taken from the top surface of the elevating handwheel. It is graduated in 1-mil increments for a total of 50 mils. The handwheel is also equipped with an indicator that points to each number on the handwheel as it is turned. Once the gunner has the major reading from the screw plate scale, he will then get the minor reading by looking at the number at which the indicator is pointing. Both portions of the elevation reading are recorded by placing a slash (/) mark between the two portions.

(e) An elevation reading is valid only on one T&E mechanism. If the same data is placed on another T&E mechanism using the same weapon, the data may be inaccurate. The number of threads exposed on the T&E must remain the same both when obtaining and recording data. If the number of exposed threads is changed in any manner, the firing will be off target. For example, when a gun is freed to engage targets in the secondary sector, the data will be correct if the gunner ensures that the same amount of threads is exposed when he returns to his primary sector of fire.

(f) To ensure that the data is correct, the gunner should fire and adjust his weapon.

(5) The T&E manipulation exercise gives the gunner practice and the instructor a tool to evaluate the gunner's progress (Figure 5-9, page 5-12). The exercise is conducted in two stages. Both stages require the coach to give directions and the gunner to respond. In the first stage, the coach

positions himself about 10 paces to the front of the gun. He then directs the gunner to manipulate the weapon in certain directions. He indicates the direction by the use of hand signals. The gunner responds by manipulating the T&E mechanism with his left hand. The gunner must keep his eyes on the coach; at no time during this exercise is he permitted to look at the T&E mechanism. The coach must be very attentive during the first stage because the gunner will be manipulating using the elevating handwheel and the traversing handwheel. The second stage is done in the same way except the gunner must make bold changes in elevation and deflection. The exercise continues until the instructor is satisfied that the gunner can manipulate the weapon by T&E without looking at the device. This exercise can also be conducted using the basic MG target. The gunner will be shifted from one selected target to another. The coach must observe all movements of the gunner during this training.

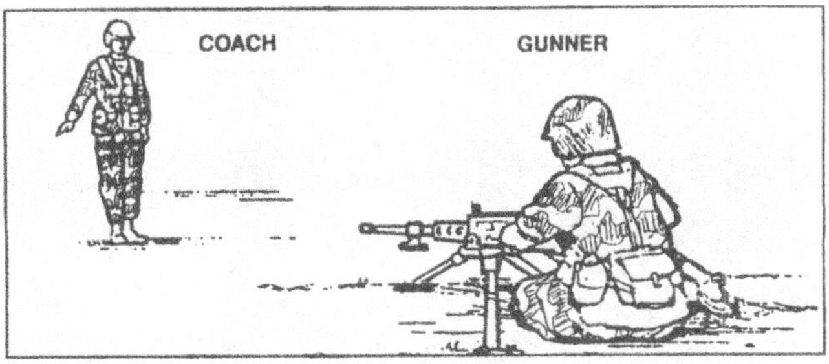

Figure 5-9. Manipulation exercise.

5-8. RANGE DETERMINATION

Range determination is the process of estimating the distance to a target from a gunner's position. The ability of the gunner to get the range to, sight on, and destroy a target is the realism of combat. Under combat conditions, ranges are seldom known in advance; therefore the effectiveness of fire depends largely upon the accuracy and speed of the gunner in determining range. Some methods of determining range are estimating by eye (Table 5-1, page 5-15), firing the gun, measuring range from a map or aerial photograph, stepping off the distance, or securing information from other units. Ranges are determined to the nearest 100 meters for machine gun firing. In combat, the most commonly used methods are estimating by eye and firing the gun. There is also a method used for measuring lateral distance.

a. The two techniques of eye estimation are the 100-meter unit of measure method and the appearance of objects method.

(1) When using the 100-meter unit of measure method, the gunner must be able to visualize what 100 meters looks like on the ground. With this distance in mind, the gunner can mentally determine the number of 100-meter units between his position and the target. The accuracy of this method is limited to 500 meters or less, and it requires constant practice (Figure 5-10).

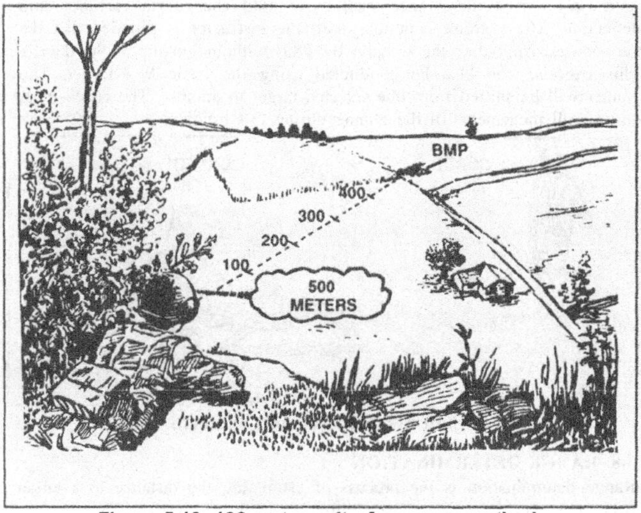

Figure 5-10. 100-meter unit of measure method, less than 500 meters.

(2) For targets that appear to be more than 500 meters, the gunner must modify this technique. The gunner selects what he thinks is the halfway point between the target and his position. He then mentally counts the number of 100-meter units to the halfway point and doubles it. This method of range determination is not accurate beyond 1,000 meters (Figure 5-11, page 5-14).

(3) Some terrain affects the appearance of 100-meter units of measure. When the terrain slopes upward toward the target, 100 meters appears longer than on level terrain. It appears shorter on downward

sloping terrain. The gunner must consider these two factors when using the 100-meter unit of measure method.

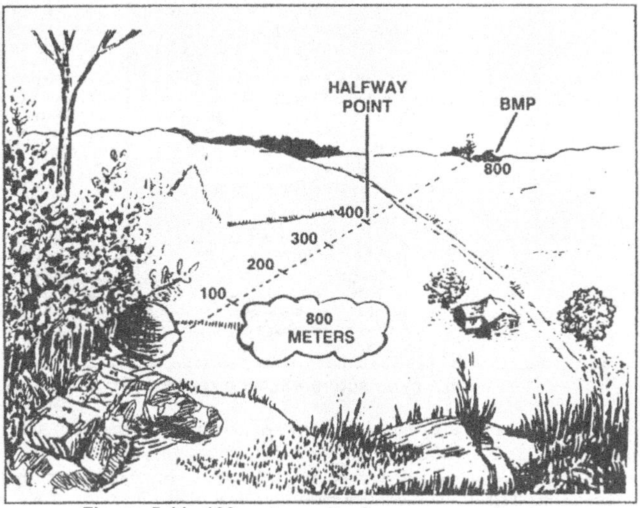

Figure 5-11. 100-meter unit of measure method, more than 500 meters.

(4) The appearance of objects method may be used if the gunner is unable to use the 100-meter unit of measure method because of terrain. To use this method, the gunner learns through practice how familiar objects look at various known ranges. This can be achieved by studying the appearance of a man standing 100 meters away. The gunner must then fix the appearance of the man firmly in his mind to include the size and details of his uniform and equipment. Next, he studies the same man at the same distance in the kneeling and prone positions. This procedure is used at 200, 300, 400, and 500 meters. By comparing the appearance of the man at these known ranges, he can establish a series of mental images that will help him determine range on unfamiliar terrain out to 500 meters. This training could also be conducted to familiarize the gunner with other objects, such as weapons and vehicles, at various ranges.

FACTOR	APPEARS NEARER (Range is underestimated when)—	APPEARS MORE DISTANT (Range is overestimated when)—
Target visibility.	Most of the target is visible and offers a clear outline.	Only a small part of the target is seen or target is small in relation to its surroundings.
Terrain, or position of the observer.	Looking across a depression, most of which is hidden from view. Looking down from high ground. Looking down a straight, open road or along a railroad track. Looking over uniform surfaces such as water, snow, desert, or grain fields.	Looking across a depression, all of which is visible. Looking from low toward high ground. When vision is narrowly confined as in streets, draws, or forest trails.
Light and atmosphere	In bright light or the sun is shining from behind the observer. The target is in sharp contrast with the background, or is silhouetted by reason of size,shape, or color,or is seen in the clear atmosphere of high altitudes.	In poor light such as dawn and dusk, in rain, snow, fog, or when the sun is in the observer's eyes. The target blends into the background or terrain.

Table 5-1. Factors affecting range estimation by eye.

b. Firing the gun is another method of determining range. In this method, the gunner opens fire on the target at the estimated range and moves the center of the beaten zone into the center base of the target by means of the T&E handwheels. He resets the sight so the new line of aim is at the center base of the target and notes the range setting on the rear sight. This range setting may apply only to this gun. When the ground in the vicinity of the target does not permit observation of the strike of the rounds, or when surprise fire on the target is desired, fire is adjusted on a point that offers observation and is known to be the same range as the target. The gunner then lays his gun on the target when ordered. When moving into position occupied by other units, range cards prepared by those units can furnish valuable range information on targets, suspected targets, and various terrain features. When the tactical situation and time permit, range may be determined by pacing off the distance.

c. Lateral distance measure is a method that the gunner may use to determine the distance from one target to another from left to right or right to left. When the gun is mounted on the M3 tripod, width can be measured by aiming on a point and manipulating the traversing handwheel, counting the clicks from one point to another point of aim. Each click equals one meter at 1,000 meters or one-half meter at 500 meters. This method is accurate but time-consuming. The finger measurement method is not a method of range determination; it is a method of measuring the lateral distance (in fingers or mils) between two points. To measure the distance in fingers between a reference point and a target, extend the arm with palm outward, the fingers cupped, and elbow locked. Close one eye, raise the index finger, and sight along its edge, placing the edge of the finger along the flank of the target or reference point (Figure 5-12). The remaining space is then filled in by raising fingers until the space is covered. The measurement is then stated as being one or more fingers or so many mils, depending on the number of fingers used (Figure 5-13).

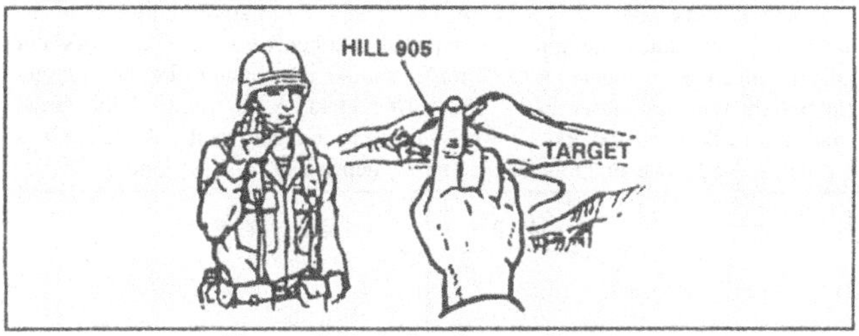

Figure 5-12. Index finger aligned.

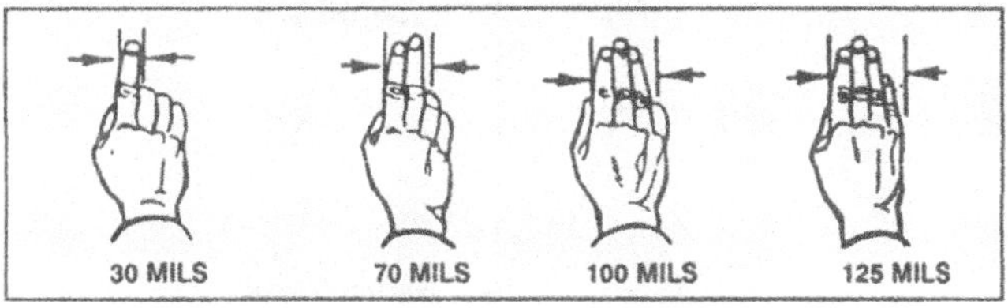

Figure 5-13. Mil/finger relationships.

5-9. OBSERVATION AND ADJUSTMENT OF FIRE

The purpose of observation and adjustment of fire practice is to teach the adjustment of fire by observing the strike of the bullets and the flight of the tracers, or by frequent re-laying on the target using sights.

a. Observation is used when firing on the 10-meter range because the impact of the round is visible on the target. When firing at greater distances, the strike of the round on the ground may cause dust to rise that is visible to the gunner; however, during wet weather, the strike cannot always be seen. In this case, use tracer ammunition that allows the gunner or crew to note the strike of the burst in relation to the target.

b. Adjustments on the target can be made using the mil relation; that is, one click of traversing or elevating handwheel moves the strike of the round one-half inch on the target at 10 meters. When firing on field targets, adjustment is made by moving the burst into the target. One click of traverse will move the strike of the round one-half meter at 500 meters, or one meter at 1,000 meters (Figure 5-14). However, the distance one click of elevation will move the strike of the round depends on the range to the target and the slope of the ground. The gunner determines the number of mils necessary to move the center of the strike into the target, and he manipulates the gun the required number of mils. This does not require the use of sights. For example, should the gunner fire on a target at 500 meters and observe the strike 10 meters to the right of the target and short about 50 meters, he would traverse the gun to the left 20 clicks (mils) and add one or more clicks (mils), depending on the slope.

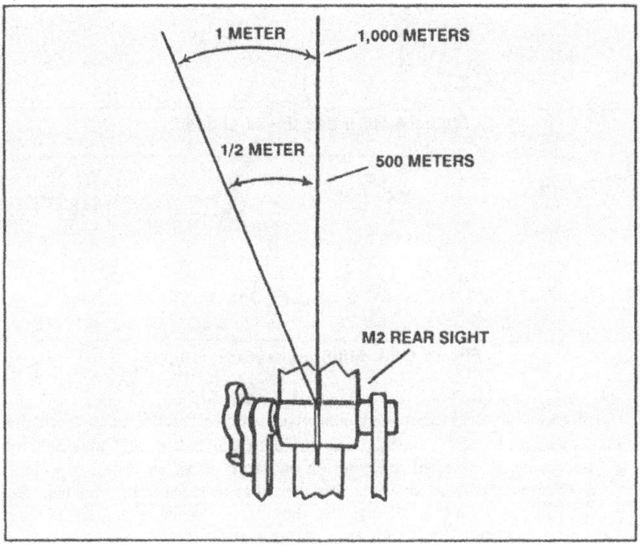

Figure 5-14. Mil relation.

c. The gunner may use the adjusted aiming point method to adjust the fire. In this method, the gunner must use his sights. He selects an aiming point that will place the next burst on target. For example, when the gunner fires on a target at 500 meters and estimates that the rounds impacted 20 meters short and 10 meters to the right, he would rapidly select an aiming point about 20 meters beyond the target and 10 meters to the left of it and lay on that aiming point and fire (Figure 5-15).

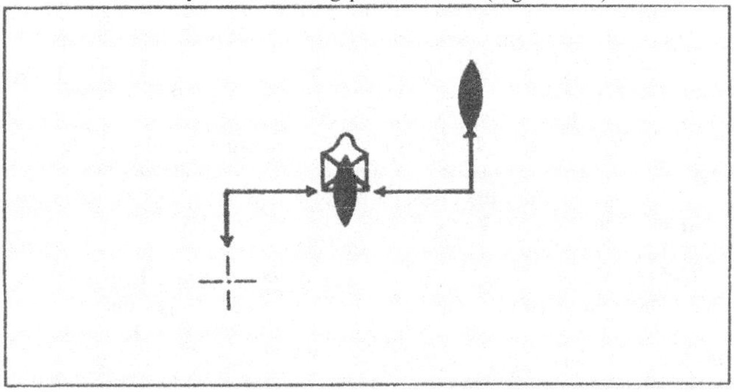

Figure 5-15. Adjusted aiming point method of fire adjustment.

5-10. FIRE COMMANDS

Fire commands are technical instructions issued by a leader to enable the unit or crew to accomplish a desired fire mission. Fire commands have been standardized for infantry direct fire weapons, and they follow the same sequence. There are two types – initial fire commands, issued to engage a target; and subsequent fire commands, which are issued to adjust fire, change the rate of fire, interrupt fire, shift fire to a new target, or to terminate the alert. A correct fire command is one that is as brief as clarity permits and yet includes all the elements necessary for the accomplishment of the fire mission. It is given in the proper sequence, transmitted clearly at a rate that permits receipt and application of instructions without confusion.

a. **Elements of the Initial Fire Command.** There are six essential elements of the initial fire command for the machine gun, which are given or implied by using one or more of the methods of control. During training, the gun crew repeats each element as it is given. This is done to avoid confusion and to train the crew to think and act in the proper sequence. The six elements are the alert, direction, description, range, method of fire, and the command to open fire.

(1) *Alert.* This element brings the crew to a state of readiness to receive further instructions. Once alerted, the gunner ensures the gun is loaded. The assistant gunner continuously checks with the leader for orders or instructions and passes them on to the gunner. The oral alert is announced as FIRE MISSION. At this command, the gunners are alerted that a target has been detected and fire may be delivered upon it. When the leader announces the alert, such as FIRE MISSION, both gun crews react to the alert. If only a certain gun is to engage, the leader announces NUMBER 1 (or 2). The other crew follows the fire mission, loads, and lays on the target to take up the fire, if required.

(2) *Direction.* This element indicates the general direction to the target and may be given in one or a combination of the following ways:

(a) The leader gives the direction orally to the target in relation to the position of the gun(s). For example, FRONT, RIGHT FRONT, LEFT FRONT.

(b) The leader can designate a small or obscure target by pointing with his arm and hand or aiming the machine gun. When pointing with his arm and hand, a man standing behind him should be able to look over his shoulder and sight along his arm and index finger to locate the target. When a gun has been aimed at a target, a soldier looking through the sights should be able to see the target.

(c) Tracer ammunition is a quick and sure method of designating a target that is not clearly visible. When using this method, the leader should first give the general direction to direct the gun crew's attention to the desired area. To minimize the loss of surprise when using tracer ammunition, the leader does not fire until he has given all the elements of the fire command except the command to fire. The leader may use his individual weapon or fire one or more bursts from the machine gun. The firing of the tracer(s) then becomes the last element of the fire command and is the signal to open fire. For example:

FIRE MISSION.
FRONT.
BUNKER.
WATCH MY TRACER(S).
SLOW (or SINGLE SHOT).

The leader fires his individual weapon or a machine gun at the enemy bunker, then his gun crew(s) opens fire.

(d) Another method of designating obscure targets is by using easily recognizable reference points. Prominent terrain features and man-made objects make good reference points. All leaders and members of the crew(s) must be familiar with the terrain features and the terminology used to describe them. The general direction to the reference point should be given. For example:

FIRE MISSION, NUMBER 2.
FRONT.
REFERENCE: LONE PINE TREE.
TARGET: TRUCK.

Sometimes a target must be designated by using successive reference points. For example:

FIRE MISSION, NUMBER 1.
RIGHT FRONT.
REFERENCE: RED-ROOFED HOUSE, LEFT TO HAYSTACK, LEFT TO BARN.
TARGET: MACHINE GUN.

Finger measurements can be used to direct the gun crew's attention to the right or left of reference points. For example:

FIRE MISSION.
LEFT FRONT.
REFERENCE: CROSSROAD. RIGHT FOUR FINGERS.
TARGET: LINE OF TROOPS.

When the guns are mounted on tripods, lateral distance from reference points can be accurately announced. When gunners are firing the tripod-mounted gun, lateral distance is assumed to be in mils unless otherwise indicated, so the word "mils" is not necessary. For example:

FIRE MISSION.
FRONT.
REFERENCE: KNOCKED-OUT TANK. LEFT FOUR ZERO.
TARGET: COLUMN OF TROOPS.

(3) *Description.* The target description is used to create a picture of the target in the minds of the gun crew. The gun crew must know the type of target they are to engage to properly apply their fire. The leader should describe it briefly but accurately. For example:

Dismounted enemy personnel .... TROOPS
Automatic weapons .... MACHINE GUN
Armored vehicles .... TANK
Artillery or antitank weapon .... ANTITANK
Airplanes or helicopters .... AIRCRAFT

If the target is obvious, no description is necessary. Finger measurements or mil measurements can be used to designate the width of a linear target when the flanks cannot be pinpointed.

(4) *Range.* The range to the target is given so the gun crew knows how far to look for the target and immediately knows what range setting to place on the rear sight. Range is determined and announced in meters. Since the meter is the standard unit of range measurement, the word "meters" is not announced. With machine guns, the range is determined and announced in even hundreds and thousands. For example: THREE HUNDRED, ONE THOUSAND, ONE ONE HUNDRED. This element may be omitted when the gunners can obviously determine the range; however, it is desirable in some situations to announce the range.

(5) *Method of fire.* This element includes manipulation and rate of fire.

(a) Manipulation is used to prescribe the class of fire with respect to the gun. It is announced FIXED, TRAVERSE, SEARCH, TRAVERSE AND SEARCH, SWINGING TRAVERSE, or FREE GUN.

(b) To control the rate of fire, the gunner may use single shot, slow, rapid, or cyclic.

- Single shot. Place the gun in the single-shot mode and engage the target with aimed shots. The MG is accurate out to 1,500 meters.
- Slow fire. Slow fire consists of less than 40 rounds per minute, in bursts of five to seven rounds, fired at 10- to 15-second intervals.
- Rapid fire. Rapid fire consists of more than 40 rounds per minute, fired in bursts of five to seven rounds, at 5- to 10-second intervals.
- Cyclic fire. Cyclic fire is when the weapon fires 450 to 550 rounds per minute.

(6) *Command to open fire.* If surprise fire is not desired, the command FIRE is given without pause. It is often important that machine gun fire be withheld for maximum effect of surprise fire. To ensure this, the leader may preface the command to commence firing with the words AT MY COMMAND or AT MY SIGNAL. When the gunner(s) is ready to engage the target, he reports UP to the assistant gunner(s) who signals READY to the leader. For example:

FIRE MISSION.
FRONT.
TROOPS.
AT MY COMMAND. (Pause until crew members are ready and fire is desired.)
FIRE (or appropriate arm-and-hand signal).

When the leader makes a mistake in the initial fire command, he corrects it by announcing CORRECTION, and then gives the corrected element(s). For example:

FIRE MISSION.
FRONT.
TROOPS.
FIVE HUNDRED.
CORRECTION.
SIX HUNDRED.
TRAVERSE.
AT MY COMMAND.

When the leader makes an error in the subsequent fire command, he may correct it by announcing CORRECTION, and then repeating the entire subsequent fire command. For example:

LEFT FIVE, DROP ONE.
CORRECTION
LEFT FIVE, DROP ONE ZERO.

b. **Subsequent Fire Commands.** If the gunner fails to adjust his fire on the target, the leader must promptly correct him by announcing or signaling the desired changes. When changes are given, the gunner makes the required corrections and continues to engage the target without further command. When firing under the control of a leader, the assistant gunner checks with the leader for instructions, which he passes on to the gunner. Changes in the rate of fire are given orally and by arm-and-hand signals. To interrupt firing, the leader announces CEASE FIRE or gives a signal to cease fire. The gun crew(s) remains on the alert and firing can be resumed on the same target by announcing FIRE. To terminate the alert, the leader announces CEASE FIRE, END OF MISSION.

5-11. CREW EXERCISES

The purpose of crew exercise is to develop precision, speed, skill, and teamwork in examining equipment, placing the gun into action, and taking it out of action. In crew exercise, precision must be stressed. Once that is attained, speed, skill, and teamwork will follow. Duties are rotated during crew exercise to allow each member of the gun crew to become familiar with all the duties. During crew exercise, all oral or visual signals are repeated. When the fire command is completed, the gunner will give the assistant gunner an UP. The assistant gunner will extend his hand and arm into the air in the direction of the leader (to indicate READY) and announce, UP. With the M3 mount, the crew must consist of at least four men, including the leader. There is no designated crew in the TOE for a dismounted caliber .50 MG. The following paragraphs are only suggestions for the breakdown of equipment and member designation that may be established by the commander.

a. **Crew Equipment.** In addition to individual arms and equipment, crew members carry the following equipment for the tripod-mounted machine gun:

CREW MEMBER	SUGGESTED MINIMUM EQUIPMENT
Squad or crew leader.	Binoculars, compass, one box of ammunition.
No. 1 Assistant gunner.	Tripod.
No. 2 Gunner.	Receiver, T&E mechanism attached, and headspace and timing gauge.
No. 3 Ammunition bearer.	Barrel, barrel cover, and box of ammunition.

b. **Form for Crew Exercise.** The crew leader commands, FORM FOR CREW DRILL.

(1) *Positions with equipment.* The crew forms in column, facing the crew leader with five paces between men (Figure 5-16, page 5-24). They are in the following order: assistant gunner, gunner, and ammunition bearer. When the crew members reach their correct positions, they assume the prone position with equipment arranged as follows:

(a) No. 1: Tripod to his left, trail legs to the rear, front leg uppermost.

(b) No. 2: Receiver across his front, backplate to the right, retracting slide handle uppermost.

(c) No. 3: Barrel to his right, muzzle to the rear, ammunition box to his left front with latch to the right (latch to the front for the new box).

(d) Other members, if present: Ammunition boxes in front, one foot apart, latches to the right (front).

(e) Crew leader: Ammunition box to his right as he faces the crew, latch to the right (front).

(2) *Rotation of duties.* Duties are rotated to ensure that each member learns and is capable of performing the duties of the other members.

(a) The command to rotate all personnel is, FALL OUT LEADER. At this command, each member of the crew rises, moves forward, and assumes a new duty. The crew leader becomes the ammunition bearer. The assistant gunner moves forward and becomes the crew leader. The gunner moves forward and becomes the assistant gunner. The ammunition bearer moves forward and becomes the gunner.

(b) If the leader is not changed, the command, FALL OUT ASSISTANT GUNNER, is given. At this command, the crew members rise, the gunner becomes the assistant gunner, the ammunition bearer becomes the gunner, and the assistant gunner becomes the ammunition bearer. When the crew members have assumed their new position, they call out their new duties in order, ASSISTANT GUNNER, GUNNER, AMMUNITION BEARER.

NOTE: An additional crew exercise, which the crew maybe required to practice, is the setting of headspace and timing. These procedures are outlined in paragraph 3-6.

Figure 5-16. Crew formed in column.

c. Inspection of Equipment Before Firing. When the crew is formed with equipment, the command is, INSPECT EQUIPMENT BEFORE FIRING. At this command, the crew proceeds as follows:

(1) The assistant gunner inspects the tripod M3 mount to ensure that:

(a) The indexing levers and clamps on the front and trail legs function properly, and the legs are in the short (low) position.

(b) The front leg and trail legs are closely folded, and the front leg clamp is hand tight.

(c) The sleeve lock latch and pintle lock release cam are in working order, and the pintle lock release cam is down.

(d) The pintle bushing is free from dirt and burrs.

(2) The gunner inspects the receiver group to ensure that:

(a) The barrel support and breech bearing are free of dirt.

(b) The gun pintle is free of dirt.

(c) The feed mechanism and bolt switch are properly assembled to feed from left (soldier raises cover for proper inspection).

(d) The striker projects through the aperture in the face of bolt (soldier closes the cover).

(e) The rear sight is set at 1,000 yards (900 meters) windage zero.

(f) The T&E mechanism is securely attached to the receiver.

(g) The traversing handwheel is centered.

(h) The elevating screws are equally exposed (about 2 inches) above and below elevating handwheel.

(i) The backplate is latched and locked in place.

(j) The bolt latch release is locked in the down position by the bolt latch release lock.

(3) The ammunition bearer inspects the barrel and ammunition box to ensure that:

(a) The barrel is clear.

(b) The barrel carrier assembly is securely attached to the barrel.

(c) The barrel threads are free of dirt.

(d) The metallic links are clean (soldier opens ammunition box).

(e) The belt is properly loaded and placed in box with the double-looped end up.

(f) Dummy ammunition is used during crew exercise, and no live ammunition is present.

(g) The box is closed and latched.

(4) When the ammunition bearer completes his inspection, he moves to the gunner's position with the barrel in his right hand and ammunition box in his left hand. With the aid of the gunner, he screws the barrel into the barrel extension (Figure 5-17, page 5-26). The headspace and timing adjustment is made. The ammunition bearer remains on the left and on line with the gunner.

(5) The crew leader examines his ammunition as described in paragraph (3).

(6) At the completion of the inspection, a report is rendered as follows:

(a) The ammunition bearer reports: AMMUNITION CORRECT (or any deficiencies).

(b) The gunner reports: GUN AND AMMUNITION CORRECT (or any deficiencies).

(c) The assistant gunner reports: ALL CORRECT (or any deficiencies).

Figure 5-17. Screwing the barrel into the barrel extension.

d. **Placement of the Gun Into Action.** To place the gun into action, the crew leader commands and signals, GUN TO BE MOUNTED HERE (pointing to the position where the gun is to be mounted), FRONT (pointing in the direction of fire), ACTION (vigorously pumping his fist in the direction of the designated gun position).

(1) At the command or signal ACTION, the assistant gunner grasps the left trail leg near the center with his left hand. Springing to his feet and grasping the tripod head with his right hand, he lifts the tripod across the front of his body with the front leg up, and carries the tripod to the desired location. Upon arrival at the position, he places the trail leg pointing upward. Steadying the tripod with his left hand on the front leg, he loosens the front leg clamp with his right hand (Figure 5-18), positions the front leg with his left hand, and tightens the front leg clamp with his right hand. With his right hand on the tripod head, he slides his left hand down on the left trail leg and with a snapping motion, pulls the left leg (to the left), engaging the sleeve latch (Figure 5-19). He then aligns the tripod for direction, drops the mount to the ground, stamps the right and left trail shoes with his right or left foot, and assumes the prone position behind the mount.

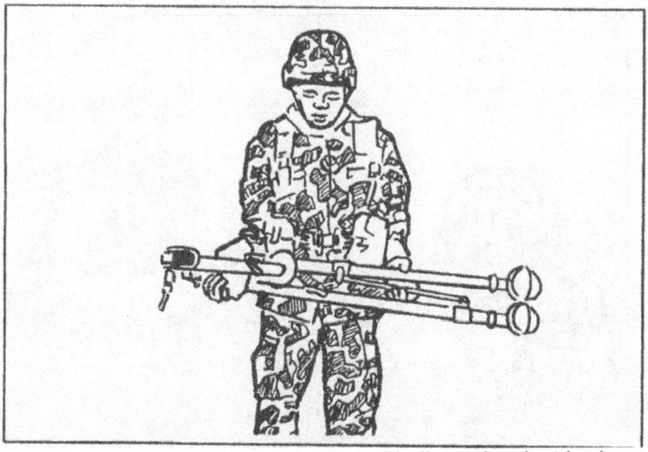

Figure 5-18. Assistant gunner-(No.1) opening the tripod.

Figure 5-19. Emplacing the mount.

(2) The gunner and ammunition bearer move together. When the tripod is nearly mounted, the gunner and ammunition bearer spring to their feet. The gunner places both hands on the spade grips, the ammunition bearer grasp the ammunition box in his left hand and the gun carrying handle with this right hand. Both men lift the gun from the ground and move rapidly to the gun position. Then the ammunition bearer places the ammunition box on line and in front of the leader's ammunition box. The gunner, assisted by the ammunition bearer, inserts the pintle into the pintle bushing. The ammunition bearer stamps the front shoe into the ground with his left foot (Figure 5-20), releases the carrying handle, lowers his right hand, and lifts the gun pintle lock release cam. When the gun pintle is fully seated, he presses down the pintle lock release cam with his right hand, turns to his left, and returns to his original position.

Figure 5-20. Mounting the gun.

(3) The assistant gunner holds the left spade grip with his left hand. With his right hand, he lowers the T&E mechanism to the traversing bar, ensuring that the traversing handwheel is to the left and the lock lever is to the rear. He then assumes a semiprone position to the left of the gun with his feet to the rear and his head on line with the feedway. He unlatches and raises the cover of the ammunition box, removes the ammunition belt,

inserts the double-looped end into the feedway, and taps the feedway with his right hand to ensure it is closed.

(4) With his right hand palm-up, the gunner grasps the retracting slide handle, vigorously jerks it to the rear, and releases it. The gun is half-loaded (Figure 5-21). He then assumes a prone position directly behind the gun with his legs spread and heels down. His right hand lightly grasps the right spade grip with the thumb in position to depress the trigger. His left hand is on the elevating handwheel (palm down) with his thumb near the traversing slide lock lever. He rests on his left elbow with his head as close as possible to the rear sight (Figure 5-22, page 5-30). He then gives the assistant gunner an UP. The assistant gunner announces UP and extends his hand and arm into the air in the direction of the leader.

Figure 5-21. Gun mounted, gunner half-loading the gun.

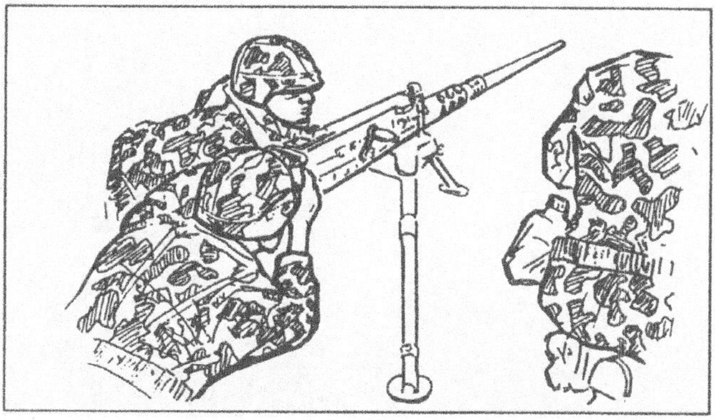

Figure 5-22. Crew ready for action.

e. **Removal of the Gun From Action.** To take the gun out of action, the command is OUT OF ACTION.

(1) The gunner raises the cover with his left hand, the assistant gunner lifts the ammunition out of the feedway, replaces the ammunition in the box, and closes and latches the box. The gunner closes the cover, pulls the retracting slide handle to the rear with right hand (palm up) and releases it, presses the trigger with his right thumb, and loosens the traversing slide lock with his left hand. He rises to his feet, grasping both spade grips. At this time, the ammunition bearer will arrive at the gun position. With his left hand, the ammunition bearer grasps the ammunition box and places his right foot on the front leg shoe. With his right hand, he reaches down and lifts up the pintle lock release cam and grasps the carrying handle (Figure 5-23).

(2) The gunner and ammunition bearer lift the gun from the tripod, turn right, and carry the gun back to their original position. They set the gun down with the muzzle to the left and the retracting slide handle up. The gunner pulls the retracting slide handle to the rear and aligns the lug on the barrel locking spring with the 3/8-inch hole in the right side of the receiver. The ammunition bearer unscrews the barrel from the receiver, picks up the ammunition box and barrel, moves five paces to the rear, places the barrel to his right with the muzzle to the rear and the ammunition box to his left, and assumes the prone position.

Figure 5-23. Crew taking the gun out of action.

(3) The gunner rotates the bolt latch release lock to the right and releases the bolt latch. He checks his sights to ensure they are properly set.

(4) The assistant gunner assumes a kneeling position behind the tripod and releases the sleeve lock latch with his right hand. He rises to his feet, grasps the front leg with his left hand, and rotates the tripod to a vertical position on the trail legs. With his right hand, he loosens the front leg clamp, folds down the front leg, then tightens the clamp. With his right hand on the tripod head, he rotates the tripod on the right trail leg and releases the sleeve latch. Then he folds the left trail leg against the right with his left hand (Figure 5-24, page 5-32). Holding the tripod head with his right hand, trail legs with his left, he lifts the tripod across the front of his body with the front leg up. He turns to the right and returns to his original position. At this time, the crew leader picks up his ammunition box and faces the crew. The assistant gunner places the tripod on the ground, assumes a prone position to the right of the tripod, and announces UP.

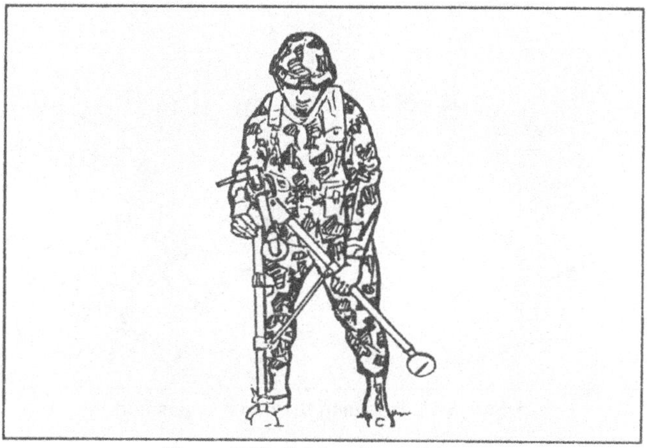

Figure 5-24. Folding the trail legs.

f. **Duties of The Crew.** To hand-carry the gun and equipment, the command is SECURE EQUIPMENT, FOLLOW ME. At this command,–

(1) The crew leader carries his ammunition box in his right hand.

(2) The assistant gunner carries the tripod over either shoulder.

(3) The gunner carries the receiver on either shoulder with the spade grips to the front.

(4) The ammunition bearer carries his ammunition box in his left hand and the barrel in his right hand with the muzzle to the rear (or on his left shoulder with the muzzle to the front).

g. Relocation of Tripod-Mounted Gun. When the gun is mounted on the tripod, it can be moved for short distances by dragging or by a two- or three-man carry. (In the latter, the men should move in step to make carrying easier.)

(1) *Dragging.* The gun is dragged when there is limited cover, or when the situation requires the gun to be moved in this manner. The gunner and assistant gunner drag the mounted gun to the desired position (Figure 5-25).

(2) *Two-man carry* With the gunner on the right and assistant gunner on the left, each grasps the front leg with his forward hand and a trail leg with the other hand, just above the traversing bar (Figure 5-26).

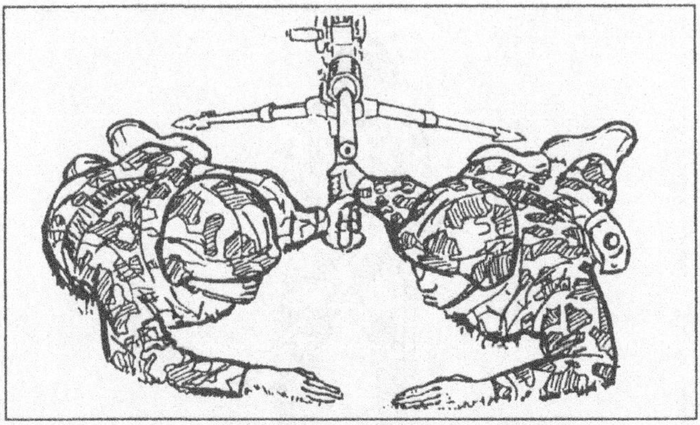

Figure 5-25. Dragging the gun into position.

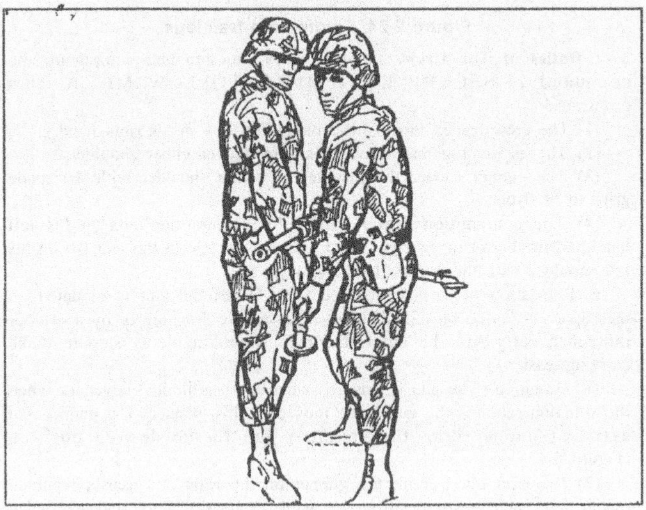

Figure 5-26. Two-man carry.

(3) *Three-man carry*. When the barrel is hot, the gunner gets behind the tripod with a trail leg in each hand. With the assistant gunner on the left and the ammunition bearer on the right, each grasps the carrying handle. In addition, the assistant gunner carries the ammunition in his left hand (Figure 5-27). When the barrel is cool, the ammunition bearer and the assistant gunner each grasp the front leg (Figure 5-28).

Figure 5-27. Three-man carry (hot barrel).

Figure 5-28. Three-man carry (cola barrel).

NOTE: Carrying the gun by the barrel may cause damage to the barrel support and the barrel extension.

h. **Movement of the Gun to Other Mounts.** With the mount prepared to receive the gun, the cradle of the mount is placed in a horizontal position. To move the gun to the mount, the gunner carries the right spade grip in his left hand and a box of ammunition in his right. The assistant gunner grasps the carrying handle with his left hand and a box of ammunition in his right hand. When they get to the mount, the gunner and assistant gunner place their ammunition boxes near the mount. The gunner removes the rear mounting (gun-locking) pin with his right hand. The assistant gunner removes the front mounting (gun-locking) pin with his right hand. They place the gun on the mount. The gunner aligns the holes in the rear mounting lugs of the receiver with the rear mounting bracket and inserts the rear mounting pin. The assistant gunner aligns the front mounting hole in the front of the receiver with the front mounting bracket and inserts the front mounting pin. (For use of the sideplate trigger with the M63 mount, see TM 9-1005-213-10.)

5-12. MACHINE GUN FUNDAMENTAL SKILLS TEST

The machine gun fundamental skills test should be held periodically to ensure that proficiency with the MG is maintained by all crewmen. It consists of 10 fundamental skills; all tasks are hands-on (Figure 5-29). The test should be given prior to range firing on a go/no-go basis.

TASK	GO	NO-GO
1. Clear the MG.	✓	
2. Disassemble the MG.	✓	
3. Assemble the MG.	✓	
4. Do a function check.	✓	
5. Maintain the MG.	✓	
6. Set the headspace on the MG.	✓	
7. Set the timing on the MG.	✓	
8. Load the MG.	✓	
9. Perform immediate action on the MG.	✓	
10. Unload the MG.	✓	

Figure 5-29. The MG fundamental skills test.

Section III. BASIC MARKSMANSHIP

This phase of training is designed to allow the gunner to apply the fundamentals of marksmanship that he learned earlier. During this training, the gunner is introduced to the basic machine gun target (see Appendix C), procedures for both the 10-meter and field fire ranges, and how to acquire targets. He also fires practice and qualification.

5-13. CONCEPT OF ZEROING/TARGETING

The concept of zero is very simple; it is no more than the best way to adjust the sights of the weapon so the point of aim of the sights and the point of impact of the rounds are the same at any given range. A properly zeroed M2 gives the gunner the highest probability of hit for most combat targets with the least adjustment to the point of aim. There are three methods of zeroing/targeting used with the .50 caliber MG.

a. Ten-meter zero is the basic and the most common method of zeroing the M2 MG. Once zeroed on a 10-meter range using the standard machine gun target, the weapon is ready for field fire. As other weapons, the sight on the M2 must also be set at an initial start point (Figure 5-30). The initial sight setting for field zero is basically the same; except the range setting during field zero will depend on the range to the target, and it is always 1,000 yards for 10 meters.

(1) *Set elevation.* Raise the rear sight by lifting straight up until it snaps into the upright position. Adjust the range scale to 1,000 yards by rotating the elevation screw knob in the necessary direction. (Clockwise moves the scale up; counterclockwise moves the scale down.)

(2) *Set windage.* Rotate the windage knob until the zero index mark on the base rear sight is aligned with the index mark on the top of the receiver. (Clockwise moves the windage scale to the left; counterclockwise moves it to the right.)

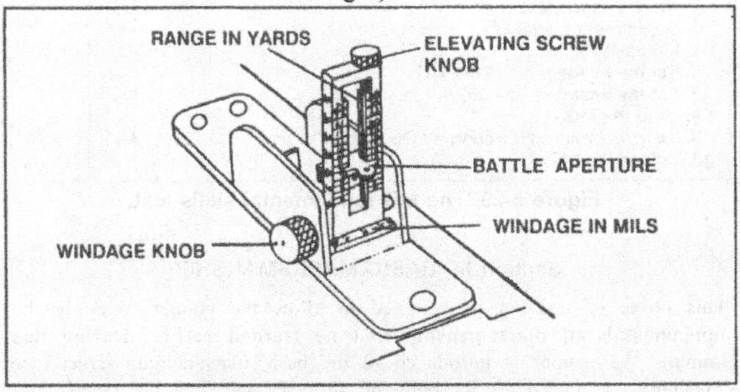

Figure 5-30. Rear sight setting.

(3) *Obtain proper sight picture.* Obtain the proper sight picture by looking through the zero aperture and centering the front sight blade in it. Once the sight alignment is obtained, place this combination on the center base of the selected target (Figure 5-31).

Figure 5-31. Proper sight picture.

b. Field zeroing/targeting is an expedient method of obtaining a zero when a 10-meter zero cannot be conducted. When preparing to field zero, make sure the M2 is mounted securely on the M3 tripod, make sure the T&E is working properly, and finally, know the distance to your zero target. The only difference in initial sight setting for field zero is range setting on the scale. The gunner must also remember that the range scale on the M2 is indicated in yards. Therefore, in order to get as close to the target as possible, you may have to convert the meters to the target into yards so you can set the range on the rear sight. Conversion of meters to yards is accomplished by multiplying the number of meters by 1.094. For example, 600 meters x 1.094 = 656.4 yards; the gunner would set his range scale at 650.

c. The AN/TVS 5 is an effective night fire aid. By using this device, the gunner can observe the area and detect and engage any suitable target. But, as usual, the device is only as good as its zero; the zeroing procedure requires practice to become proficient.

(1) *Mount the mounting bracket.* To mount the bracket, the gunner must ensure that the rear sight is in the down position. He then releases the catch on the left side of the top cover assembly and raises the cover to the upright position. He pushes the mounting bracket over the breech of the machine gun and slides the bracket rearward until it stops (Figure 5-32).

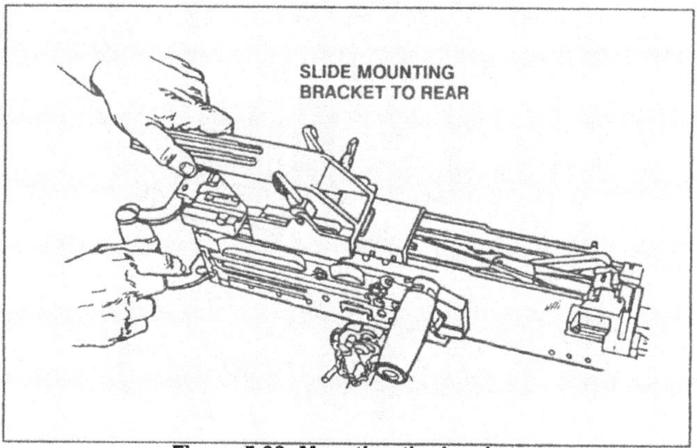

Figure 5-32. Mounting the bracket.

He pushes the three locking cams to secure the bracket (Figure 5-33) and closes the top cover assembly.

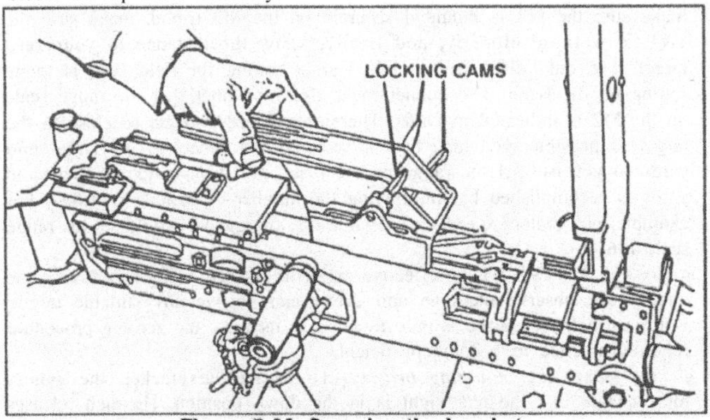

Figure 5-33. Securing the bracket.

(2) *Install the sight on the mount.* To install the sight, the gunner must align the scribe lines on the sight with the scribe lines on the bracket. He places the sight in the groove at the top of the bracket and tightens the lever screw to secure the sight to the bracket (Figure 5-34).

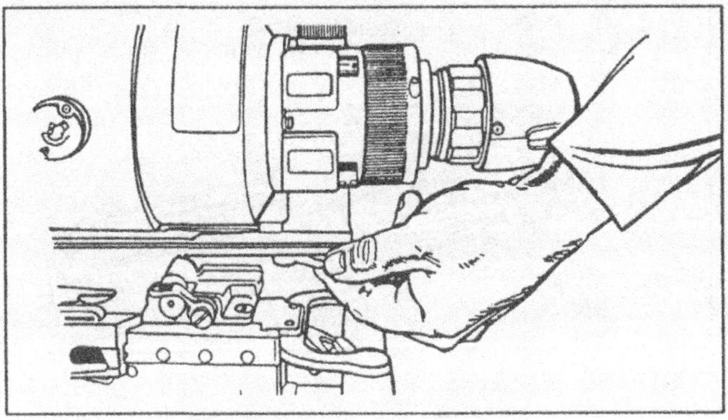

Figure 5-34. Securing the sight.

(3) *Zero the AN/TVS-5.* A gunner must zero the MG with its NVD on it before he can effectively engage targets with the AN/TVS-5. The zeroing of this device to the MG must be done at 50 meters. An M16A1, A2, or Canadian bull zero target may be used. The point of impact for either target must be 9.8 centimeters below the target aiming point. To accomplish this task, place the reticle aiming point on the target aiming point (Figure 5-35).

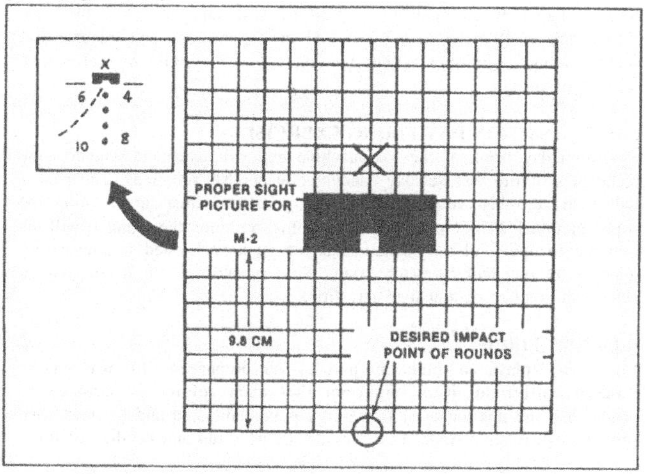

Figure 5-35. Desired impact point for the AN/TVS-5.

(a) Fire enough single-shot rounds to obtain a good shot group. Locate the center of the shot group. Adjust the AN/TVS-5 azimuth and elevation knobs until the shot group impacts 9.8 centimeters below the target aiming point. Move the weapon so that the reticle aiming point is again on the target aiming point. Repeat this process until the desired point of impact is obtained.

(b) When adjusting, move the azimuth or elevation adjustment actuator one click to move the strike of the round .5 inch at 50 meters. One click of adjustment will move the reticle about one square on the Canadian bull-type target.

NOTE: The lens cover with the peephole may be required to prevent scope washout from the muzzle flash.

(4) *Dismount the sight.* The gunner dismounts the sight from the bracket first by loosening the lever screws. Then he lifts the sight from the bracket. He must then dismount the bracket in the reverse mounting procedure.

5-14. TEN-METER FIRING EXERCISE

The purpose of 10-meter firing is to develop skills in the delivery of initial burst on target. When conducted properly, it will train gunners in the basic skills, such as 10-meter zero, controlled bursts, traverse, and traverse and search firing techniques. (See Appendix C for details on setup and conduct of firing.)

5-15. TRANSITION DAY FIRING EXERCISE

Transition day firing of the M2 machine gun will teach the gunner some techniques of fire that he may encounter in combat situations. The gunner will field zero his weapon and engage point and area targets from the tripod-mounted firing position. Within this training, the gunner will be required to apply all the fundamentals of gunnery learned in preparatory gunnery training and 10-meter firing. (See Appendix C for details on the setup and conduct of transition day fire.)

5-16. NBC FIRING

Since NBC plays an important part in our preparation for war on the modern battlefield, it is important that each soldier is prepared to accomplish the mission even if the area is contaminated and he must wear protective gear. (See Appendix C for details on setup and conduct of fire.)

5-17. NIGHT FIRE EXERCISE

The night fire exercise gives the soldiers the practical application of engaging targets using the AN/TVS-5 at night or during limited visibility. (See Appendix C for details on setup and conduct of fire.)

Section IV. ADVANCED GUNNERY

After firing 10-meter, day, NBC, and night, gunners need practice in applying what they have learned. They also need experience in engaging targets that depict realistic enemy formations. Advanced gunner exercises provide this experience in mounted, mounted NBC, and predetermined firing exercises.

5-18. OBJECTIVES

The objectives of the advanced gunnery phase are to prepare the gunners for combat. During this phase, training should be directed toward –

- Tracking and leading.
- Mounted target engagement techniques.
- Mounted NBC engagements.
- Predetermined fire techniques.

5-19. TRACKING AND LEADING EXERCISES

The gunner normally completes instruction in firing at stationary targets before he receives instruction in firing at moving targets. The technique of engaging a moving target differs from that of engaging a stationary target. The gun must be aimed ahead of the moving target a sufficient distance to cause the bullet and target to arrive at the aiming point at the same time. The distance is measured in target lengths. One target length as seen by the gunner is one lead. Leads are measured from the center of mass. The lead depends upon range, speed, and direction of movement of the target. To hit the target, the gunner aims at a point ahead of the target equal to the estimated number of leads, maintains the lead by tracking the target (manipulating the gun at the same angular speed as that of the target), and then fires. Fire is adjusted by observation of strike/tracer (Figure 5-36).

Figure 5-36. Lead technique.

a. **Tracking.** Tracking consists of maintaining correct alignment of the sights (with or without a lead) on a moving target by moving the gun at the same angular speed as that of the target.

(1) The gunner is required to aim at a prescribed point (center base to take advantage of the beaten zone) on the target and maintain that aim during uniform movement of the target. As instruction progresses, speeds used should differ for successive runs of the target. The speeds the 10-meter targets should move to represent speeds at various ranges are shown in Table 5-2.

TARGETS SPEEDS IN MILES PER HOUR	TARGET SPEEDS IN INCHES PER SECONDS CORRESPONDING TO–		
	300 M	500 M	900 M
7 1/2	6	4	2
15	12	8	4
30	24	15	9

Table 5-2. 10-meter moving target.

(2) The target handler must have practice in moving the target silhouette across the background at the varying speeds.

b. **Leading.** Mathematical computation or use of voluminous lead tables to obtain exact leads on a moving target are impractical in combat. The simple lead table shown in Table 5-3 gives the amount of lead necessary to hit a target moving at right angles (90 degrees) to direction to hit at speeds and ranges indicated.

SPEED IN MILES PER HOUR	RANGE OF TARGET		
	300 M	500 M	900 M
15	1/2 target length	1 target length	2 target lengths

Table 5-3. Lead table.

(1) The gunner must correct the lead as conditions change. If the target speed is 7 1/2 miles per hour, the amount of lead is half that shown in the table; at 30 miles per hour, double that shown, The angle at which the target is moving also alters the amount of lead taken. If the angle between line of fire and line of travel of the target is less than 45 degrees, use half the lead shown in the table.

(2) For targets moving directly toward the gun, the point of aim is placed on the center or the lower edge of the target, when possible. For targets moving directly away from the gun, the point of aim is placed on the center or upper edge of the target. Too much lead is better than too

little because the target runs into the fire; also, the observation of strike is easier. Intelligent use of the lead table includes immediate application of fire with estimated lead followed by necessary corrections based upon observation of strike/tracer.

c. **Tracking and Leading.** Combined tracking and leading exercises at 10 meters are used to gain proficiency in tracking the target. The gunner is required to repeat the tracking exercise while using a designated lead to simulate firing when his sights are properly aligned. As a further exercise in tracking and leading, the gunner may be required to track and lead moving targets at greater ranges. A vehicle can be run at right angles to the line of aim at ranges between 500 and 1,000 yards, and at varying speeds, averaging 15 miles per hour (Figure 5-37).

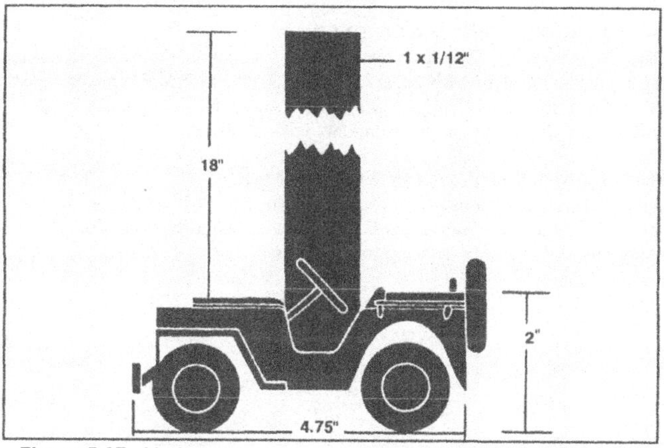

Figure 5-37. Aiming target used in tracking and leading exercise.

d. **Conducting the Lead Exercise (10-Meter).** The gunner is required to take a position at the gun, swing through the target's silhouette, and aim at a point ahead of the target equal to the prescribed lead from the center of mass. The gunner then directs the target handler to move the marking silhouette until the center of the target is at the point of aim. He repeats this procedure three times for each target lead announced. The target handler places his marking silhouette on the blank background, traces around it, and holds it in place for the gunner to aim, using the prescribed leads. Following the gunner's instructions, he moves the marking silhouette until the gunner commands HOLD. He then places a pencil dot at this point and returns the silhouette to the original position. This procedure is followed until the gunner has completed three tries for each target lead announced. The three pencil dots for each target should fit within a one-centimeter circle. The exercise should be conducted for varying left and right leads.

5-20. MOUNTED FIRING EXERCISE

The purpose of the mounted firing exercise is to teach the gunner techniques of firing the M2 MG from a mounted platform and to develop the gunner's ability to fire the M2 with it mounted on its primary carrier. (See Appendix C for details on the setup and conduct of fire.)

5-21. MOUNTED NBC FIRING EXERCISE

The probability of fighting mounted in a tactical environment that has been contaminated by NBC agents is very likely; therefore, gunners must be trained to engage targets while in a mounted NBC posture. (See Appendix C for details on the setup and conduct of fire.)

5-22. PREDETERMINED FIRING EXERCISE

The predetermined firing exercises are designed to instruct the gunners on preparing and using range cards during any visibility conditions. (See Appendix C for details on setup and conduct of fire.)

CHAPTER 6

COMBAT TECHNIQUES OF FIRE

Technique of fire is the method of delivering and controlling fire Each member of the machine gun crew must be trained in standard methods of applying fire, either as a crewmember or a gunner, and must perform his assigned task automatically and effectively. The simplest and most effective technique of delivering fire with the machine gun mounted on its ground or vehicular mount is to align the sights of the gun on the target and fire; this is called direct laying. At times, techniques of fire other than direct laying are more appropriate and effective. When delivering overhead fire or fire from position defilade, the gunner must use the appropriate technique.

Section I. FUNDAMENTALS

Before the machine gun can be employed to the best advantage using any firing technique, certain fundamentals must be understood and applied. These include:

- Characteristics of fire.
- Classes of fire.
- Fire control.
- Target types and methods of engagement.
- Overhead fire.
- Methods of laying the gun.
- Firing from position defilade.
- Final protective fires.

6-1. CHARACTERISTICS OF FIRE

The gunner's knowledge of his machine gun is not complete until he learns something of the action and effect of the projectiles when fired. This section discusses various characteristics of machine gun fire, including trajectory, cone of fire, and the beaten zone.

a. **Trajectory.** The trajectory is the curved path of the projectile in its flight from the muzzle of the weapon to its impact. The major factors that influence the trajectory are the velocity of the round, gravity, rotation of the round, and resistance of the air. The farther the round travels, the greater the curve of the trajectory. The highest point of the trajectory is called the maximum ordinate. This is a point approximately two-thirds of the distance from the gun to the target. The maximum ordinate increases as the range increases (Figure 6-1).

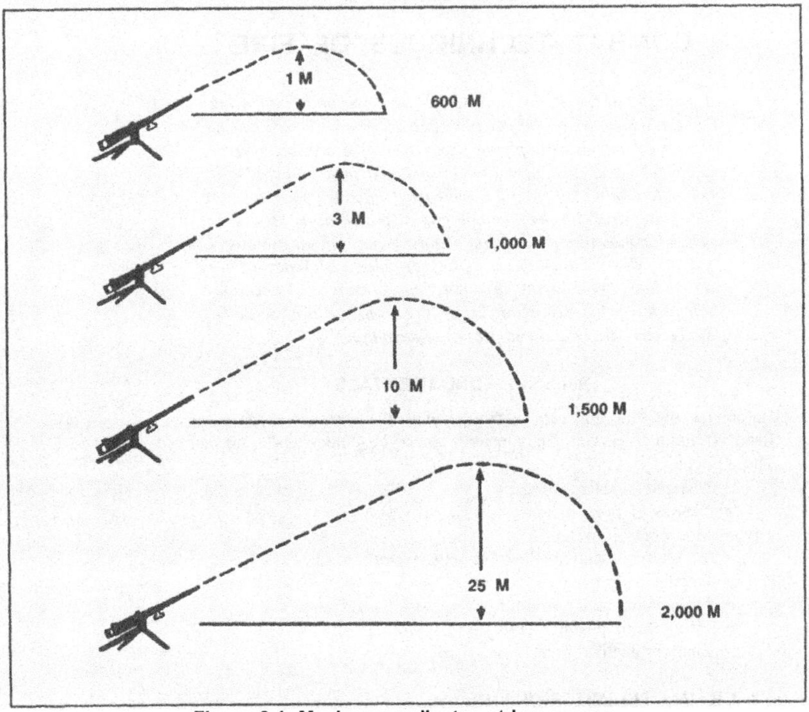

Figure 6-1. Maximum ordinates at key ranges.

b. **Cone of Fire.** When the weapon is fired automatically in bursts, all the rounds do not follow the same path. This is due to the vibrations of the gun and mount, variations in ammunition, and atmospheric conditions, which cause the rounds to follow a slightly different trajectory. This group of trajectories formed by a single burst is called the cone of fire (Figure 6-2).

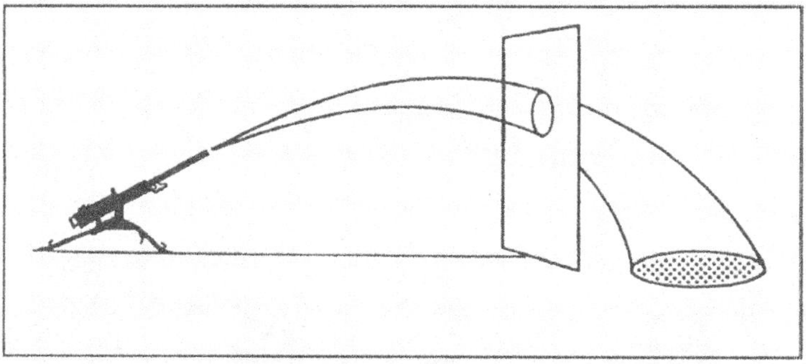

Figure 6-2. Cone of fire.

c. **Beaten Zone.** The beaten zone is an elliptical pattern formed by the cone of fire as it strikes the ground. The beaten zone is always about 2 meters in width.

(1) *Effect of range.* As the range to the target increases, the beaten zone becomes shorter and wider.

(2) *Effect of terrain.* The length of the beaten zone for any given range will vary according to the slope of the ground. On rising ground, the beaten zone becomes shorter but remains the same width. On ground that slopes away from the gun, the beaten zone becomes longer but remains the same width.

6-2. CLASSES OF FIRE

Machine gun fire is classified with respect to the ground (Figure 6-3, page 6-4), the target (Figure 6-4, page 6-5), and the gun (Figure 6-5, page 6-7).

a. **Classes of Fire with Respect to the Ground.**

(1) *Plunging fire.* Fire in which the angle of fall of the rounds (with reference to the slope of the ground) is such that the danger space is confined to the beaten zone, and the length of the beaten zone is materially shortened. Plunging fire is obtained when firing from high ground to low ground, when firing from low ground to high ground, and when firing at long ranges.

(2) *Grazing fire.* Grazing fire is fire in which the center of the cone of fire does not rise more than one meter above the ground. When firing over level or uniformly sloping terrain, the maximum extent of grazing fire obtainable is about 700 meters.

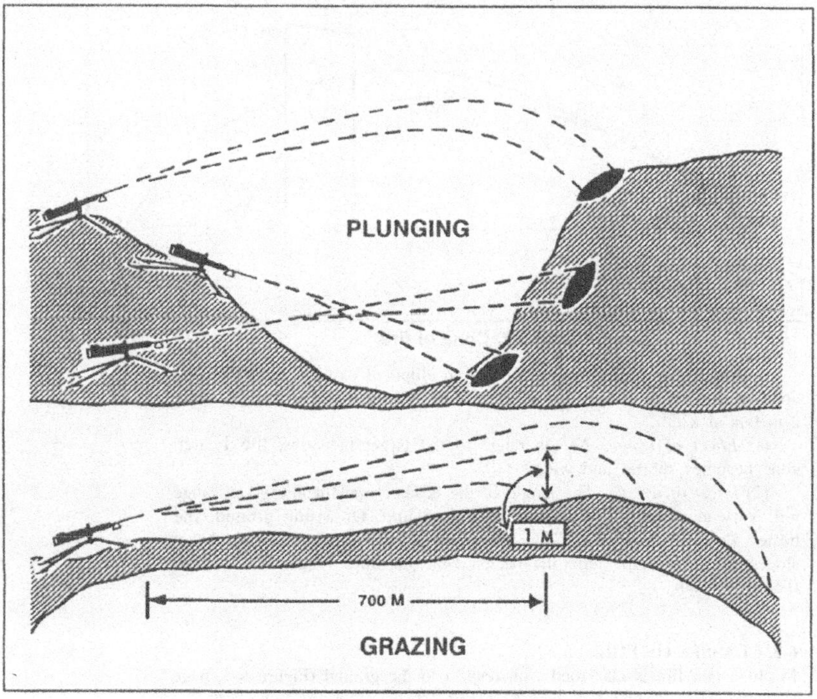

Figure 6-3. Classes of fire with respect to the ground.

b. **Classes of Fire with Respect to the Target.**

(1) *Frontal.* The long axis of the beaten zone is at a right angle to the long axis of the target.

(2) *Flanking.* Fire is delivered against the flank of a target.

(3) *Oblique.* The long axis of the beaten zone is at an angle (but not a right angle) to the long axis of the target.

(4) *Enfilade.* The long axis of the beaten zone coincides or nearly coincides with the long axis of the target. This class of fire is either frontal or flanking. It is the most desirable class of fire with respect to the target because it makes maximum use of the beaten zone.

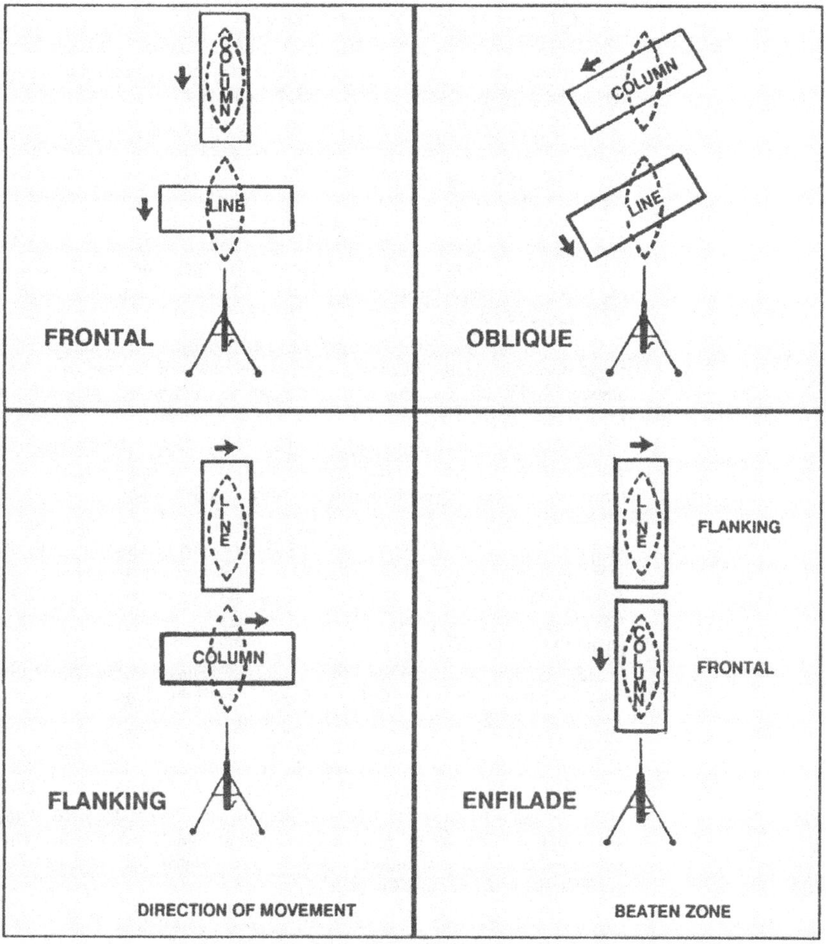

Figure 6-4. Classes of with respect to the target.

c. Classes of Fire with Respect to the Gun.

(1) *Fixed fire.* This is fire delivered on a point target with little or no manipulation needed. After the initial burst, the gunners will follow any change or movement of the target without command.

(2) *Traversing fire.* This is fire distributed against a wide target requiring successive changes in the direction of the gun. When engaging a wide target requiring traversing fire, the gunner should select successive aiming points throughout the target area. These aiming points should be close enough together to ensure adequate target coverage; however, they need not be so close as to be wasteful of ammunition by concentrating a heavy volume of fire in a small area. Two clicks on the traversing handwheel after each burst ensure coverage (2 clicks = 2 mils = constant width of beaten zone).

(3) *Searching fire.* This is fire delivered against a deep target or a target that has depth, requiring changes in elevation of the gun. The amount of elevation change depends upon the range and slope of the ground.

(4) *Traversing and searching fire.* This is fire delivered both in width and depth by changes in direction and elevation. It is employed against a target whose long axis is oblique to the direction of the fire.

(5) *Swinging traverse.* This is employed against targets that require major changes in direction but little or no change in elevation. Targets may be dense, wide, in close formations moving slowly toward or away from the gun, or vehicles or mounted troops moving across the front. The traversing slide lock lever is loosened enough to permit the gunner to swing the gun laterally.

(6) *Free gun.* This is when fire is delivered against moving targets that must be rapidly engaged with fast changes in both direction and elevation. Examples are aerial targets, vehicles, mounted troops, or infantry in relatively close formations moving rapidly toward or away from the gun position. To fire free gun on the M3 tripod mount, remove the T&E mechanism from the receiver and traversing bar and put it down. When firing swinging traverse and free gun, the weapon is normally fired at the cyclic rate of fire which is in excess of 450 rounds per minute. This consumes a lot of ammunition, and there is no beaten zone because each round seeks its own area of impact.

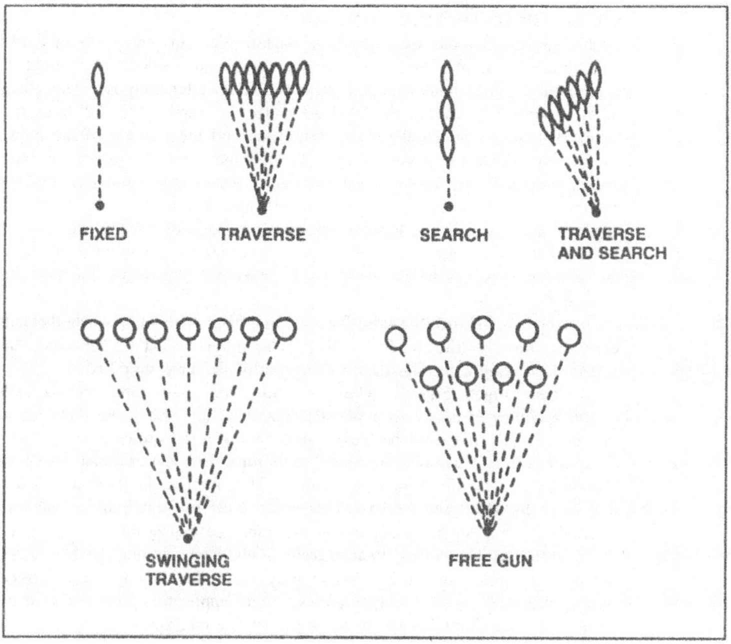

Figure 6-5. Classes of fire with respect to the gun.

Section II. FIRE CONTROL

Fire control of machine guns includes all operations connected with the preparation and actual application of effective fire on a target. It implies the-ability of the leader to-open fire at the instant he desires, to adjust the fire of the gun(s) on the target, to regulate the rate of fire, to shift from one target to another, and to cease firing. This ability to exercise proper fire control depends primarily on the discipline and the proper training of the crew. Failure to exercise fire control results in danger to friendly troops, loss of surprise, premature disclosure of positions, application of fire on unimportant targets, loss of time in adjusting fire, and waste of ammunition.

6-3. METHODS OF FIRE CONTROL

Of the several methods of controlling machine gun fire, the noise of battle will limit the use of some. Therefore, the leader must select the method or combination of methods that will best accomplish his purpose. The chain of fire control begins with the leader. He is responsible for both the technical and tactical employment of the gun and the training of the crew. He is responsible for passing on to the crewmembers all instructions and orders from his next higher leader regarding the situation and mission. He assigns sectors of fire and firing positions, designates targets to be engaged, adjusts fire, and ensures effective coverage of the targets.

a. **Oral.** Oral control is effective unless the leader is too far away from the gun crew(s), or the noise of battle makes it impossible for the gun crew(s) to hear him.

b. **Arm-and-Hand Signals.** This is an effective method when the gun crew(s) can see the leader. All crewmembers must understand the standard arm-and-hand signals used to control machine gun fire.

c. **Prearranged Signals.** These are either visual or sound signals such as pyrotechnics or blasts on a whistle. These signals should be included in the units' SOPs and must be clearly understood by all crewmembers.

d. **Personal Contact.** In many situations, the leader must move to individual crewmembers to issue orders. This method is used more than any other by small-unit leaders. The leader must use maximum cover and concealment to keep from disclosing the gun crew's position.

e. **Standing Operating Procedures.** SOPs cover actions the gun crews perform without command because they were developed and practiced during the training of the gun crews. Their application eliminates many commands and simplifies the leader's job of fire control.

6-4. TARGETS AND THEIR ENGAGEMENT

Targets presented to the machine gunners during combat will in most cases consist of enemy soldiers in various formations, which require distribution and concentration of fire. These targets have width and depth, and the application of machine gun fire is designed to completely cover the area in which the enemy is known or suspected to be. These targets may be easy to see or may be indistinct and difficult to locate.

a. When machine gun fire is under direct control of a leader, he designates the midpoint and flanks or ends of a target unless they are obvious to the gun crew(s). When a target other than a point target is engaged by two gunners, it is always divided. Each gunner applies his fire to that portion of the target corresponding to his position with relation to the other gun. Normally, each gunner engages one-half of the target;

however, gunners must be prepared to engage the entire target if necessary. Gunners continue to fire on the target until it is neutralized or until another signal is received from the leader.

b. The gunner's positions (including vehicular-mounted) should be numbered so each gunner will know which portion of a target he should engage. It should be emphasized that the positions are numbered – not the guns or gunners. To ensure that gunners react quickly and properly when they detect a target or when a target is designated by the leader, standard methods of applying fire to the various type targets are taught. These methods are the same for ground and vehicular-mounted guns. The following are the different types of targets and how they are engaged with the MG.

(1) Point targets are targets that require the use of a single aiming point. Enemy bunkers, weapon emplacements, vehicles, small groups of soldiers, and aerial targets such as helicopters or descending paratroopers are examples of point targets. A point target is engaged with fixed fire. If the target moves after the initial burst, the gun crew(s) keeps fire on the target by following its movement with the gun(s).

(2) Linear targets have sufficient width to require traversing fire and no more depth than can be effectively covered by the beaten zone. Linear targets are engaged with traversing fire.

(a) Two guns, normal division. The target is divided at the midpoint; the right gun engages the right half of the target, and the gun on the left engages the left half of the target. The point of initial lay and adjustment for both guns is at the midpoint of the target. After adjusting on the midpoint, the right gun traverses the right half of the target to include one aiming point beyond the last visible target flank and returns to the midpoint.

(b) Two guns, special division. If one portion of the target presents a greater threat than another, the target can be divided so fire is concentrated on that portion presenting the greatest threat. The special division of the target is accomplished by a subsequent fire command after firing begins. The gunners initially lay at the midpoint, regardless of the special division to be made, thus precluding confusion.

(c) One gun. A single gunner must engage the entire width of a linear target. The point of the initial lay and adjustment is on the midpoint, or that portion of the target presenting the greatest threat. The gunner traverses to either flank and then covers the remainder of the target (Figure 6-6, page 6-10).

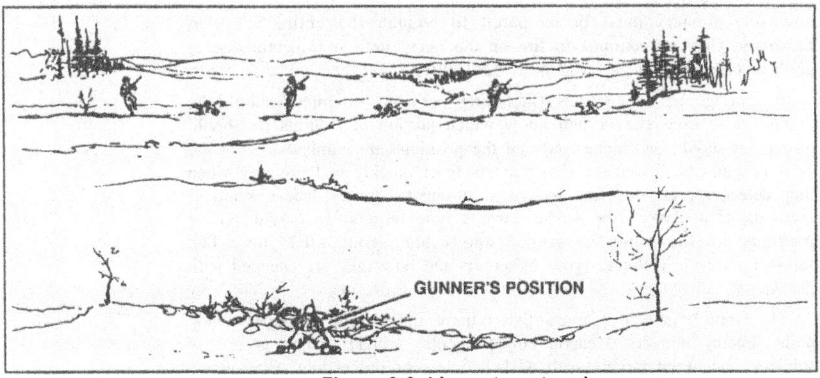

Figure 6-6. Linear target and one gun.

(3) Linear targets with depth are targets that have sufficient width to require traversing fire and depth which cannot be covered by the beaten zone. A combined change in direction and elevation (traversing and searching fire) is required to maintain effective fire on these targets (Figure 6-7). Linear targets with depth are engaged with traversing and searching fire. When range is announced, the range to the midpoint is given.

Figure 6-7. Linear target with depth.

(a) Two guns. The method of division, the point of initial lay and adjustment, and the extent of manipulation for both guns are the same as prescribed for linear targets. The gunners, however, apply enough search between each burst to ensure the center of the beaten zone is maintained at the center base of the target (Figure 6-8).

(b) One gun. A single gunner initially lays and adjusts on the midpoint of a linear target with depth unless some other portion of the target presents a greater threat. The gunner traverses and searches to the near flank, then he covers the entire target area (Figure 6-8).

(4) Deep targets have depth but very little width and can be effectively covered by searching fire (Figure 6-9, page 6-12). When the range is announced, it is given to the midpoint of the target.

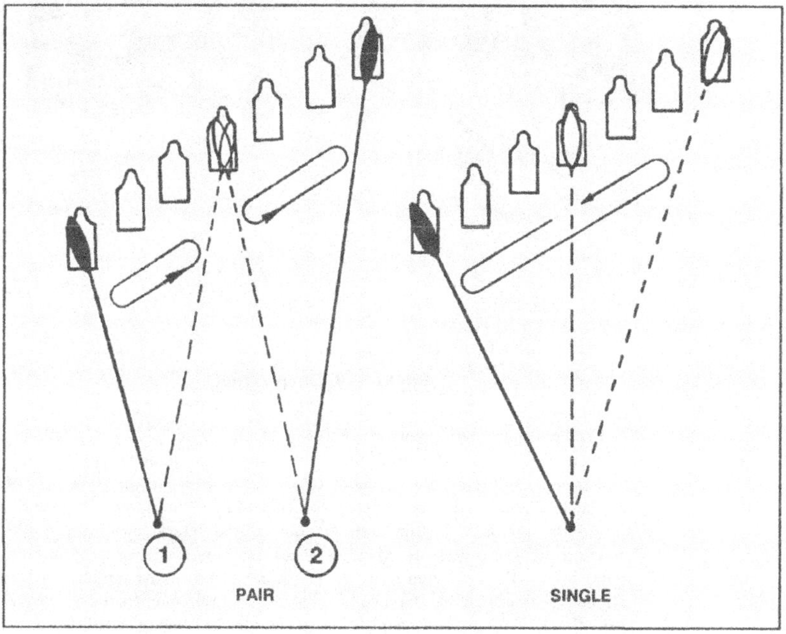

Figure 6-8. Engagement of linear targets with depth.

Figure 6-9. Deep target.

(a) Two guns. The point of initial lay of both guns is on the midpoint, which is also the point of division. Since enfilade fire is delivered, it is not necessary to adjust on the midpoint of the target because the long axis of the beaten zone will compensate for missing the midpoint. However, should the gunner's beaten zone be out of the lateral confines of the target, it will be necessary to adjust fires into the target area. After the initial bursts, the right gun searches to the near end of the target, and the left gun searches to the far end of the target. Both gunners then reverse their direction of search and return to the midpoint (Figure 6-10).

(b) One gun. A single gunner initially lays and fires at the midpoint of a deep target, unless another portion of the target presents a greater threat. The gunner immediately searches to the near end, then covers the entire target (Figure 6-10).

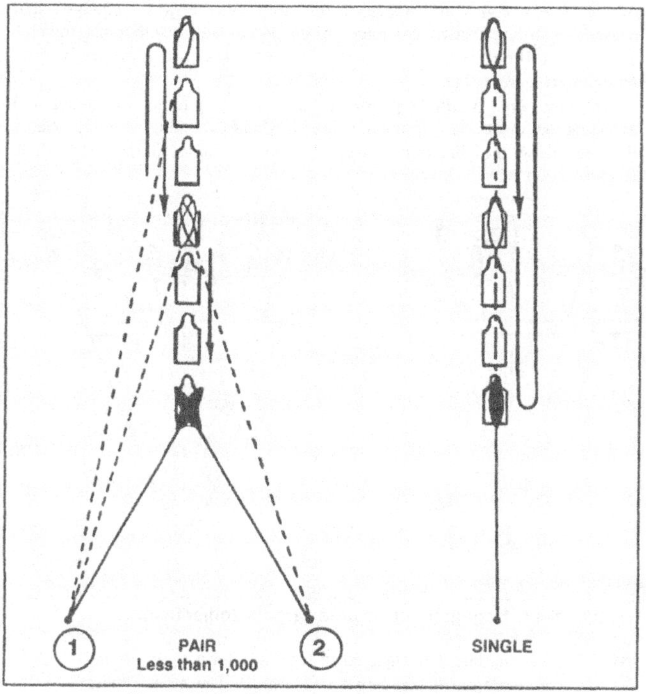

Figure 6-10. Engagement of deep targets.

(5) Area targets as discussed in this manual have considerable width and depth, and they require extensive traversing and searching fires. This type target exists when the enemy is known to be in a certain area, but his exact location is not known. A hilltop is a typical area target. The leader designates an area target by indicating to the gun crew(s) the width and depth of the target.

(a) Two guns. The target is divided at the center of mass; the right gun fires on the right half and the left gun fires on the left half. The point of initial lay and adjustment for both guns is on the center of mass. After adjusting on the center of mass, fire is distributed by determining the size of the beaten zones and applying direction and elevation changes that cause the most effective coverage of the target area. Both guns traverse and search their respective halves to the flanks, then return to the midpoint (Figure 6-11).

(b) One gun. A single gunner engages an area target by laying and adjusting on the center of mass, traversing and searching to either flank, then reversing the direction, traversing and searching to the other flank (Figure 6-11).

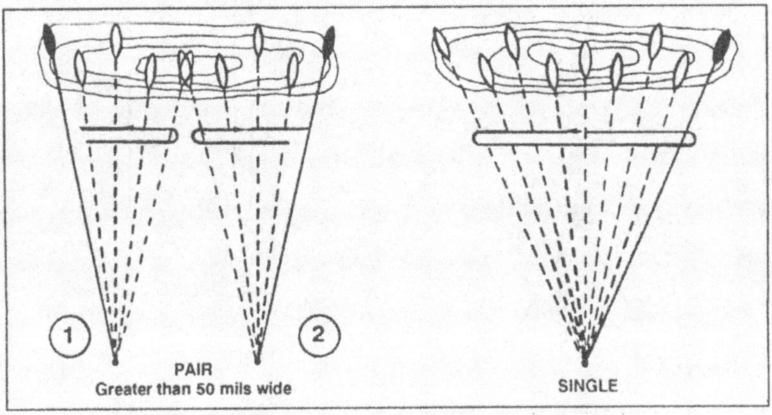

Figure 6-11. Engagement of area targets (objective).

NOTE: After the target is engaged in whatever formation it is in, the configuration of that target will change. The gunner must be trained to compensate for this change and still place effective fire on the target.

6-5. OVERHEAD FIRE

Overhead fire is fire delivered over the heads of friendly troops. A machine gun on a tripod is capable of delivering this type of fire because of the small and uniform dispersion of the cone of fire. In the attack, the use of overhead fire permits the machine gun to support the advance of rifle units. Sectors of fire allow the trainers to plan safe training while still incorporating the combat realities of overhead fires.

a. **Minimum Clearance.** The center of the cone of fire must clear the heads of the friendly troops by a prescribed distance (Figure 6-12). This distance, known as minimum clearance, is found by adding together the following elements:

- The height of a standing man, taken as 1.8 meters.
- Half the vertical dimension of the 100-percent cone of fire at the range to the troops.
- A margin of safety equal to the vertical distance which extends a 5-mil angle at the gun or 3 meters, whichever is greater.
- An additional allowance to compensate for a 15-percent error in range determination.

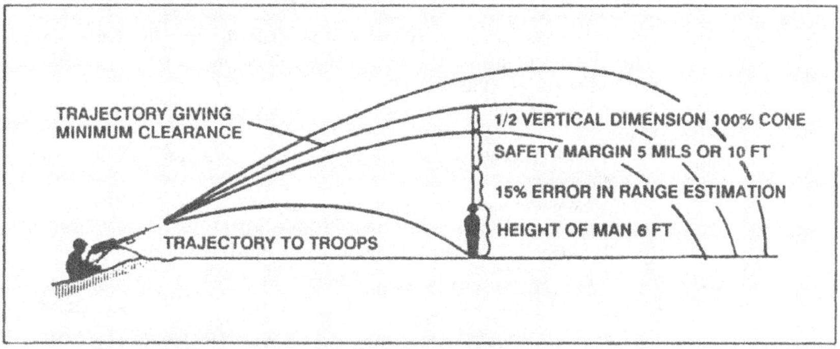

Figure 6-12. Components of minimum clearance.

b. **Safety Angles.** To obtain this minimum clearance, the gun is elevated so that the center of the cone of fire is raised from the feet-of the friendly troops to maintain clearance above their head. The amount of this elevation change is known as the safety angle. When the gun is fired from the tripod with the required safety angle, the center of impact determines the shortest range at which fire can be delivered over the heads of friendly troops. The range from the gun to the point of strike is called the corresponding range. When the ground is level or uniformly sloping between the gun and the target, the corresponding range for the safety angle used is obtained by converting the angle of elevation expressed in mils into range.

c. **Conditions.** Overhead fire is used only when the following conditions have been met:

(1) The safety limit has been determined and has been identified on the ground.

(2) The gun mount is firmly seated.

(3) Friendly troops have been notified, if at all possible, that fire is to be delivered over them.

(4) The rate of fire does not exceed 40 rounds per minute.

(5) The gun barrel is not badly worn. This condition is indicated by excessive muzzle blast.

d. **Uneven Terrain.** Level or uniformly sloping ground is seldom found in the field. This limits the use of firing tables and corresponding ranges in determining the limit of troop safety. In lieu of firing tables, a rule of thumb has been devised to give the gunner a simple method of checking for troop safety.

(1) The gunner's rule can be applied when the friendly troops are at least 350 meters in front of the gun position, and the range to the target is 850 meters or less (Figure 6-13).

(a) Lay the gun on the target with the correct sight setting to hit the target.

(b) Without disturbing the lay of the gun, set the rear sight at a range of 1,600 meters.

(c) Look through the sights and notice where the new line of aim strikes the ground. This is the limit of troop safety. When the feet of the friendly troops reach this point, fire must be lifted or shifted.

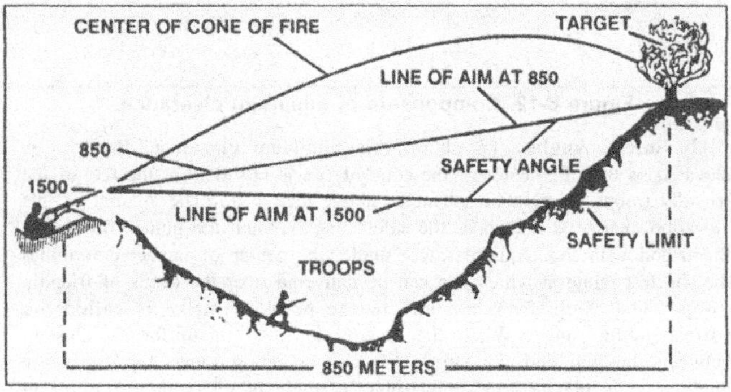

Figure 6-13. Application of gunner's rule.

(2) When the range to the target is greater than 850 meters, overhead fire should be delivered only in an emergency and then only out to a range in which either the tracers or the strike of the bullets can be seen by the gunner. In this situation the leader's rule applies (Figure 6-14).

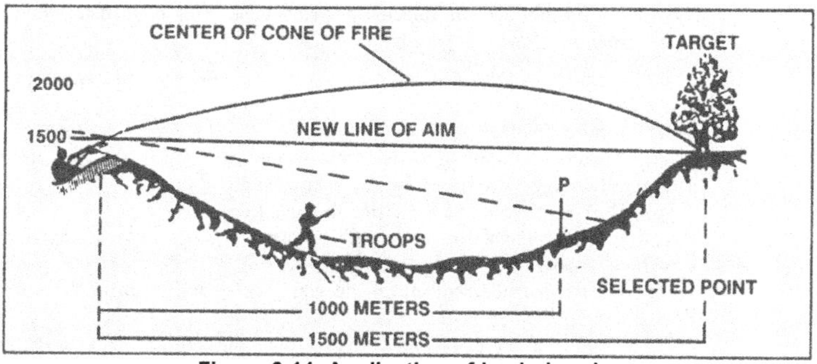

Figure 6-14. Application of leader's rule.

The platoon or section leader uses the leader's rule only when the target is greater than 850 meters. The rule is as follows:

(a) Select a point on the ground where it is believed friendly troops can advance with safety.

(b) Determine the range to this point by the most accurate means available.

(c) Lay the gun on the target with the correct sight setting to hit the target.

(d) Without disturbing the lay of the gun, set the rear sight to 1,600 meters, or the range to the target plus 500 meters, whichever is greater. Under no conditions should the sight setting be less than 1,500 meters.

(e) Note the point where the new line of aim strikes the ground.

- If it strikes at the selected point, that point marks the limit of safety.
- If it strikes short of the selected point, it is safe for troops to advance to the point where the line of aim strikes the ground and to an unknown point beyond. If it is desired to fire after friendly troops advance farther than the point where the line of aim strikes the ground, this farther point is determined by testing new selected points until the line of aim and the selected point coincide.

- If it clears the selected point, it is safe for the troops to advance to the selected point and to an unknown point beyond. If it is desired to have troops advance beyond the selected point, this farther point must be determined by testing new selected points until the line of aim and the selected point coincide. This point marks the line of safety.

e. **Precautions.** The following safety precautions *must* be observed in delivering overhead fire.

(1) Firmly emplace the tripod mount.

(2) Use depression stops to prevent the muzzle of the gun from accidentally being lowered below the safety limit.

(3) Do not deliver overhead fire through trees.

(4) Inform commanders of friendly troops when fire is to be delivered over their heads.

(5) Ensure that all members of the gun crew(s) are aware of the safety limit.

(6) Do not deliver overhead fire if the range from the gun to the target is less than 350 meters or more than 850 meters.

(7) Do not use a barrel that has excessive muzzle blast or is otherwise determined to be badly worn.

(8) Do not lay machine guns so their fire will cross at any point over the heads of friendly troops.

6-6. DEFILADE POSITIONS

To achieve maximum effectiveness, the machine gun must be employed using the technique of direct lay; however, at times it may be desirable to employ guns from defilade positions.

a. **Full Defilade.** A machine gun is in defilade when the gun and its crew are hidden from enemy ground observation by a land mass such as the crest of a hill. The position may be on the reverse side of the crest or the forward slope of the next higher ground (Figure 6-15). The gun must fire up and over the hill. Fire must be observed and adjusted by a crewmember who can observe the target from a position on a flank or to the rear of the gun (on higher ground). A defilade position allows little opportunity to engage new targets.

b. **Partial Defilade.** A machine gun is in partial defilade when a mask (usually the crest of a hill) provides the gun and gunner with some protection from enemy direct fire, but the gunner is able to engage the target using direct laying techniques. The gun is far enough up the slope so that the gunner can see the target through the sights but the lower portion of his body and lower portion of the gun are protected by the mask. Partial defilade positions are desirable when a fire mission cannot be accomplished from a defilade position.

c. **Advantages.** The gun and crew have cover and concealment from direct fire weapons. The crew has some freedom of movement in the vicinity of the gun position, and control and supply are facilitated. The characteristic smoke and flash of the gun are partially concealed from observation.

d. **Disadvantages.** Rapidly moving ground targets are not easily engaged because adjustment of fire must be made through an observer. Targets close to the mask usually cannot be engaged, and it is difficult to secure grazing fire for a final protective line.e.

e. **Position Selection.** The fire unit leader selects the location of the gun position. To select a position in partial defilade, he moves up the reverse side of the slope until he has the target in view above the mask when sighting at the height of the gunner's eye. To select a position in maximum defilade, he estimates the lowest point below the mask at which the gun can still engage the target without danger of hitting the mask.

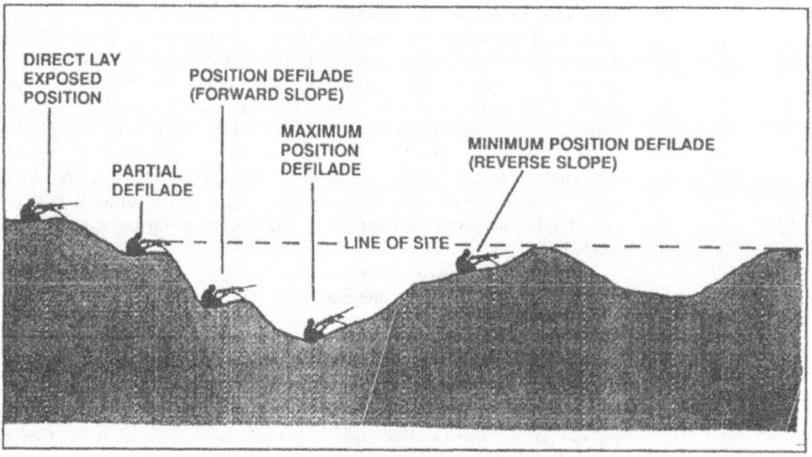

Figure 6-15. Minimum and maximum position defilade, partial defilade, and direct lay areas.

6-7. METHODS OF LAYING THE GUN FOR DEFILADE FIRING

The essential elements in engagement of a target from defilade position are direction, elevation, mask clearance, and adjustment of fire.

a. **Direction.** An observer places himself on the gun-target line in a position from which he can see the gun and the target. He aligns the gun approximately by having the gunner shift the mount. The gunner then loosens the traversing slide lock lever and, as directed by the observer, moves the gun right or left until it is aligned on the target; he then clamps it in that position. A prominent landmark, visible to the gunner through his sights, is selected as an aiming point. An aiming point on the gun-target line and at an equal or greater range than the target is desirable. However, an aiming point on the mask may be used. If the aiming point is on the gun-target line, the gun is laid on the aiming point and is thereby aligned for direction. If the aiming point is not on the gun-target line, the deflection is measured by binoculars or compass. This measured deflection is laid off with the gun.

b. **Elevation.** An aiming point visible from the gun position is selected (preferably a point at a greater range and at a higher elevation than the target) and the range to the target is determined. The leader, using binoculars, measures the vertical angle in roils from the aiming point to the base of the target. He then lays the gun on the aiming point with the sight set to hit the target. He directs the gunner to manipulate the gun through the number of mils measured. For example, in Figure 6-16, the range to the target is 1,300 meters. The angle read with the binoculars from the aiming point down to the base of the target is 12 mils. The sight is set at 1,300 meters, the gun laid on the aiming point, and the muzzle depressed 12 mils. If the aiming point is off the gun-target line, deflection in mils may be taken with the rear sight windage screw knob if it is not over 5 mils; otherwise, the deflection must be taken up on the traversing handwheel.

c. **Mask Clearance.** After the gun has been laid, determine if the entire cone of fire will clear the mask.

(1) *Visual method.* When the range to the mask is not more than 450 meters, mask clearance exists when the axis of the bore is elevated 7 mils or more above the gun-mask line. Mask clearance can be checked after the gun has been laid on the target by depressing the muzzle of the gun 2 mils and sighting along the bottom of the receiver and the barrel support. If this line of sight clears the mask, the clearance exists. Elevate 2 mils before firing.

(2) *Firing tables method.* Determine the range to the mask and obtain the corresponding angle of elevation for mask clearance from the firing

tables. The range corresponding to the angle of elevation is set on the gun sight. If the line of aim through the sight clears the mask, the clearance exists.

d. **Adjustment of Fire.** Under field conditions, even the most practical methods of laying the gun on the target quickly do not always result in the initial burst being on the target. For this reason, adjustment of fire on the target is essential. Creeping fire should be avoided. (See paragraph 6-10 for details on adjustment of fire.)

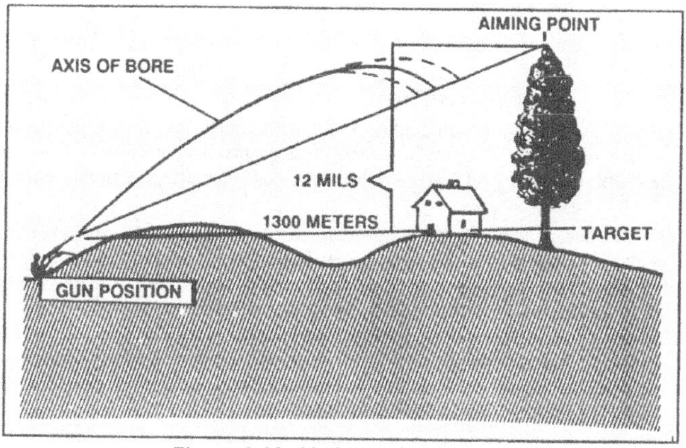

Figure 6-16. Aiming point method.

6-8. FINAL PROTECTIVE FIRES

These are types of fire that are placed on a predetermined line along which grazing fire is placed to stop an enemy assault. This fire is fixed in direction and elevation; however, a few roils of search are employed during firing to compensate for irregularities in the terrain. FPLs are always laid in using the extreme left or right of the tripod, causing the T&E to move to the extreme left or right on the traversing bar. The FPFs can be delivered in any visibility conditions. When terrain permits, final protective lines are assigned to machine guns along the forward line of troops as a part of the FPFs of the defending unit. The signal used to call for FPFs is normally prescribed in the company operation order. The authority to call for these fires may be delegated to the platoon leader of a forward rifle platoon. Final protective fires are ceased on order.

a. **Signals.** Arm-and-hand signals, voice commands, or pyrotechnic devices may be used in calling for these fires.

b. **Rates of Fire.** When firing FPFs, the rapid rate of fire is used unless it is obvious that a different rate is necessary to accomplish the mission. When engaging other preselected target areas, the rapid rate of fire is used until commanded to cease fire.

6-9. APPLICATION OF FIRE

To be effective, machine gun fire must be distributed over the entire target area. Improper distribution of fire results in gaps which allow the enemy to escape or use weapons against friendly positions without effective opposition.

a. The method of applying fire to a target is generally the same for either a single gun or a pair of guns. Direct laying is pointing the gun for direction and elevation so that the sights are aligned directly on the target. Fire is delivered in width, depth, or in a combination of the two. To distribute fire properly, the gunners must know where to aim, how to adjust their fire, and the direction to manipulate the gun. The gunner must aim, fire, and adjust on a certain point of the target. Binoculars may be used by the leader to facilitate fire adjustment.

b. The gunner ensures throughout his firing that the center of the beaten zone is maintained at the center base of the target for maximum effect from each burst of fire. When this is done, projectiles in the upper half of the cone of fire will pass through the target if it has height, and the projectiles in the lower half of the beaten zone may ricochet into the target (Figure 6-17).

c. The gunner must move his beaten zone in a certain direction over the target. The direction depends on the type of target and whether the target is engaged with a pair of guns or a single gun. When engaging targets other than point targets with a pair of guns, the targets are divided so that fire is evenly distributed throughout the target area. Fire delivered on point targets or a specific area of other target configurations is called concentrated fire.

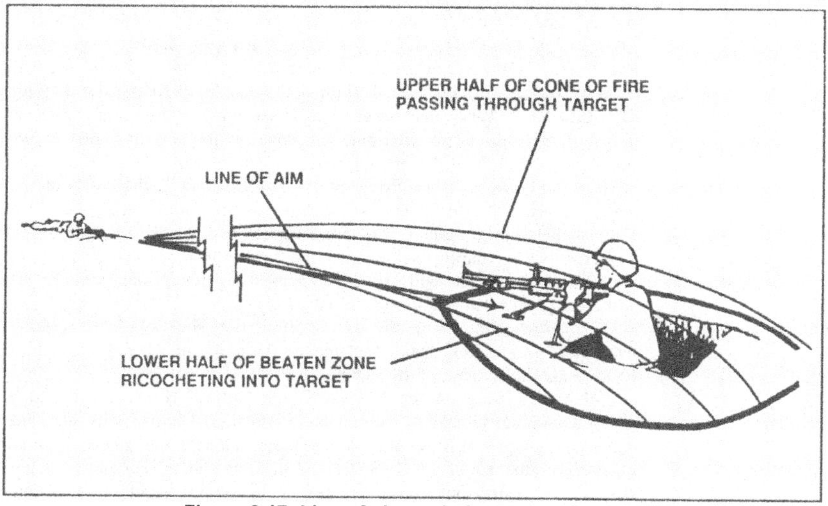

Figure 6-17. Line of aim and placement of center of beaten zone on target.

6-10. FIRE ADJUSTMENT

Machine gun fire is adjusted by observing the strike of the rounds, observing the flight of tracers, frequently re-laying the gun, or by a combination of these. Adjustment by observation of fire is the most important element of fire control if it is bold, aggressive, rapid, and continuous throughout the action.

a. The gunner is trained to observe and adjust his gun's fire without command. He is trained to anticipate the action of the enemy after the initial burst, and is prepared to shift his fire to cover any change in formation or movement of his target. If the gunner fails to accomplish this, the fire unit leader must promptly correct him by announcing or signaling subsequent fire commands. This responsibility to adjust fire continues through the chain of commands.

b. When subsequent fire commands are given, the gunner makes the required corrections and continues to engage the target without any further command to fire. If the gun is fired on the tripod mount, subsequent commands are given to make changes in direction, elevation, and the rate of fire. These changes are given orally as SHIFT RIGHT, SHIFT LEFT, ADD, or DROP. (For arm-and-hand signals see FM 21-60.) When making these announced changes,mils may be used to indicate the amount of desired shift; for example, SHIFT RIGHT 5 or SHIFT LEFT 7. When making changes in elevation, roils are not used, as it is normally difficult to determine just how high or low the center of the beaten zone is striking the ground in relation to the target.

(1) *Observation.* When firing on the 10-meter range, the strike of the bullets is visible on the target. When firing at greater distances, the strike of the bullets on the ground may cause dust to rise, which is visible to the gunner; however, during wet weather the strike cannot always be seen. In this event, the tracers will allow the gunner or crew leader to note the strike of the burst in relation to the target.

(2) *Adjustment.* Using the mil relation, one click of the traversing handwheel or elevating handwheel moves the strike of the bullet 1/2 inch on the target at a range of 10 meters.

(a) When firing on the 10-meter range, adjust by moving the shot group a required number of centimeters vertically or horizontally until the center of the group is on the aiming paster. Should the gunner's initial burst strike the target 2 centimeters to the left and 3 centimeters below the aiming paster, he adjusts his fire by traversing right 4 clicks and elevating 6 clicks before firing again.

(b) When firing on field targets, adjust by moving the burst into the target. One click (roil) on the traversing handwheel will move the strike 1/2 meter at 500 meters or 1 meter at 1,000 meters; however, the distance 1 click (roil) in the elevating handwheel will move the strike depends on the range to the target and the slope of the ground. The gunner determines the number of roils necessary to move the center of the strike into the target, and he manipulates the gun the required number of roils. This does not require the use of sights. For example, should the gunner fire on a target at 500 meters and observe the strike 10 meters to the right of the target and short about 50 meters, he would traverse the gun to the left 20 clicks (roils) and add one or more clicks (roils), depending on the slope of the ground.

(c) The gunner may use the adjusted aiming point method to adjust the fire. In this method the gunner must use his sights. He selects an aiming point that will place the next burst on the target. For example, should the gunner fire on a target at 500 meters and estimate that the strike is 20 meters short and 10 meters to the right of the target, he would rapidly select an aiming point approximately 20 meters beyond the target and 10 meters to the left of the target, lay on that aiming point, and fire.

6-11. ANTIAIRCRAFT GUNNERY

The MG can provide units with a self-defense capability against hostile low-flying, low-performance aircraft. These guns are employed in the air defense role as part of the unit's local defense. The MGs are not components of an integrated and coordinated air defense system. Unless otherwise directed, hostile aircraft within range of the gun (about 800 meters maximum effective range) should be engaged. The decision will be made by the commander. Typical targets are surveillance, reconnaissance, and liaison aircraft; troop carriers; helicopters; and drones.

a. **Engagement and Employment. The** mission is to impose maximum attrition upon the attacking enemy, such as low-flying, low-performance aircraft. Employment of MGs used for air defense is guided by the following defense design factors:

- Defense design should produce an equally balanced defense that is effective in all directions, unless a forced route of approach exists.
- Machine guns should be sited so that the maximum number of targets can be engaged, continuous fire can be delivered, and the most likely routes of approach are covered.
- Machine guns used to defend march columns should be interspersed in the convoy, with emphasis on the lead and rear elements (Figure 6-18).

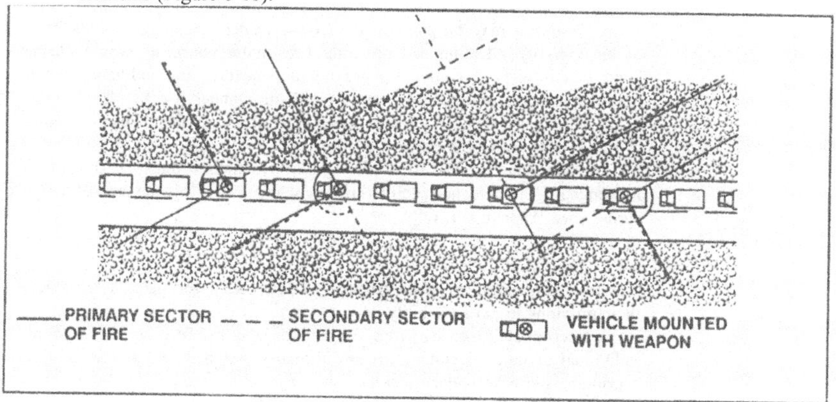

Figure 6-18. March column with four MGs (added).

b. **Target Selection and Engagement Control. These** actions depend upon visual means. The sites selected for the guns must provide maximum observation and unobstructed sectors of fire. Units furnished MGs in sufficient numbers should site them within mutual support distances of 90 to 360 meters. Each gun is assigned a primary and secondary sector of fire. Weapon crews maintain constant vigilance in their primary sectors of fire, regardless of the sector in which the guns are actually engaged.

Section III. LIMITED VISIBILITY CONDITIONS

The machine gun is provided with a stable tripod mount, M3, and a traversing and elevating mechanism. By manipulating the T&E mechanism, gun crews can record target data during good visibility and engage the same targets in poor visibility. This section provides guidance on machine gun firing techniques and terms used during limited visibility, which includes darkness, smoke, fog, rain, or snow.

6-12. DIFFICULTIES

Crewmembers encounter difficulties while defending during limited visibility, which preclude the use of many of the daylight techniques of engaging targets.

a. During limited visibility, the machine gunner's sector of responsibility cannot be observed in depth; therefore, targets are difficult or impossible to detect.

b. Visibility may be so limited that the leader cannot control the fires of his guns by selecting and directing fire on targets as he would during good visibility. Oral commands are not dependable, arm-and-hand signals may not be seen, and personal contact with the gunner is difficult.

c. At night, machine gunners have a tendency to fire indiscriminately at noises and suspected enemy locations.

To overcome these difficulties, special techniques must be developed for engaging targets and delivering preplanned fires by the use of range cards. (See Appendix E.)

6-13. TERMINOLOGY

The following terms must be familiar to MG crews for them to complete their missions in poor visibility.

a. **Sector of Fire.** An area (to be covered by fire) assigned to an individual or unit. Machine guns are normally assigned two sectors of fire, a primary and a secondary sector.

b. **Final Protective Line.** A predetermined line along which grazing fire is placed to stop an enemy assault. The FPL is fixed as to direction and elevation; however, a few roils of search are employed during firing to compensate for irregularities in the terrain. The FPL can be delivered regardless of visibility conditions. The FPL is always the inner limit of the primary sector, which is assigned close to the forward line of troops area. When terrain permits, FPLs are assigned to machine guns along the **FLOT** as a part of the final protective fires of the defending unit.

c. **Principal Direction of Fire.** A PDF is a priority direction of fire that marks a specific area assigned to a weapon. This area may extend from the gun position to the maximum effective range of the weapon and therefore is not fixed for elevation. Visible targets appearing in the PDF take priority over targets that may appear elsewhere in the sector. A PDF may be assigned to cover an area that provides good fields of fire, is a likely avenue of foot approach, or mutually supports an adjacent unit.

d. **Sector of Graze.** A wedge-shaped area formed by assigned sector limits that afford grazing fire (one meter high, maximum) from the muzzle of the weapon to the first major break in the terrain. The sector of graze is fired using swinging traverse in the primary sector of fire. It can be fired in the secondary sector in conjunction with field expedients by freeing the T&E mechanism and using the mount as a pivot. A sector of graze can be delivered regardless of the condition of visibility.

e. **Area of Graze.** This is an area, other than the sector of graze, within a sector of fire that is covered by grazing fire. Grazing fire need not be continuous from the muzzle of the weapon to the area over which grazing fire is desired.

6-14. TARGET ENGAGEMENT

A gunner's ability to detect and identify targets during limited visibility will vary, depending upon the amount of natural and artificial light and the types and numbers of sensors used. All tracer ammunition allows a gunner to more effectively engage visible targets during limited visibility; it should be used when possible. Gunners must be trained to fire low initially and adjust up when engaging targets during limited visibility. This helps them overcome the tendency to fire high during these conditions. The types of point targets machine gunners will be concerned with during limited visibility, particularly at night, are enemy automatic weapons and assaulting enemy personnel.

a. Point targets such as automatic weapons may be identified during limited visibility by their muzzle flashes. To effectively engage these targets, fire should be delivered in a heavy volume and adjusted by observing the tracer stream.

b. During the final stage of an enemy assault, machine guns normally fire at personnel on an FPL; they may be assigned a PDF. Both are considered as final protective fires and should be planned for and coordinated as such. If individual enemy soldiers are observed in the proximity of the gun position, they must be neutralized by someone other than the machine gunner (by the other crewmembers or by security forces of the supported unit). The FPFs are fired according to the order or SOP, and the machine gunner is not allowed to stop firing them except in accordance with those orders or SOP.

6-15. FIRE CONTROL

During limited visibility, the leader cannot direct the fires of his guns as effectively as with good visibility. Consequently, initiative is required of the gunners. When targets within their sectors become visible to gunners, they must engage such targets without command and continue to fire until the targets have been neutralized. Gun crews engage targets only when they can identify them, unless ordered to do otherwise. For example, if one gunner detects a target and engages it, the other gunners will observe the area in which fire is being placed. They will add their fire only if they can identify the target or are ordered to place fire in the area.

6-16. PREPLANNED FIRES

In addition to engaging appropriate visible targets, the gunner must be able to deliver preplanned fires during limited visibility. These fires are used to cover target areas of tactical significance (such as routes, avenues of approach, anticipated enemy supporting weapons positions, and probable enemy assault positions) and to establish sectors of graze and final protective lines. For maximum effect in all preplanned target areas, grazing fire should be obtained when possible.

a. **Obtaining Maximum Extent of Grazing Fire Over Level or Uniformly Sloping Terrain.** The machine gunner sets the rear sights at 700 meters; selects a point on the ground, which he determines to be at a range of about 700 meters; and lays, fires, and adjusts on this point. If the gunner cannot obtain 700 meters of grazing fire because of a major break in the ground at a range of less than 700 meters, he places the range to the break on his sight and lays, fires, and adjusts at that point.

b. **Determining the Extent of Grazing Fire on the Final Protective Line.** The extent of grazing fire on the FPL is determined using the techniques described above. Any intermediate breaks in the terrain along this line that cannot be covered by grazing fire from a gun firing along the line is considered dead space.

c. **Determining the Extent of Grazing Fire in the Sector of Graze.** The ranges to the extent of grazing fire in a sector of graze are determined by observing the terrain and by observing the tracer stream from behind or from a flank of the gun position. Normally, the extent of grazing fire within this area will be much less than on an FPL and will form an irregular pattern.

d. **Determining the Amount of Grazing Fire in an Area of Graze.** The same procedures used in paragraph 6-8a are used in determining the extent of grazing fire in an area of graze. The ranges to areas of grazing fire are determined by observing the flight of tracer ammunition from behind or from the flank of the gun position. The gunner determines the lateral extent of areas of graze by selecting and engaging successive aiming points in the area believed to afford grazing fire, using the same range setting as when determining the range to the extent of grazing fire.

6-17. NBC CONSIDERATIONS

During this phase of training, the gunner is introduced to firing the machine gun while in MOPP, keeping in mind that engagement of some targets in MOPP is a qualification requirement. Firing weapons is only part of the overall NBC training. Soldiers must first be familiar with the NBC equipment, its use, and proper wear before they progress to learning the techniques of MOPP firing. Although there is no different technique required to fire the MG, there are certain fundamentals that may be slightly impaired.

a. **Immediate Action.** Under normal conditions, a gunner should be able to clear a stoppage in two to four seconds; however, under full MOPP, this may take a few seconds longer. Dry-fire practice under these conditions is necessary to reduce time and streamline actions. When practicing with the hood/mask and gloves, care must be taken not to snag or damage the gloves or dislodge the hood/mask during movement. Trainers should apply immediate action to a variety of stoppages during dry fire until the gunners are able to instinctively do it without compromising their NBC environment.

b. **Target Detection.** Techniques and principles of target detection and target acquisition still remain valid during NBC conditions, but considerations must be made for limiting factors imposed by MOPP equipment. For example, vision is limited to what can be seen through the mask's lens/faceplate. Peripheral vision is severely restricted. The lens/faceplate may be scratched or partly fogged, thus further restricting vision. Gunners requiring corrective lenses must be issued insert lenses before training. Scanning movements may be restricted by the hood/mask. Any of these factors could adversely affect the gunner's ability to quickly and accurately detect targets. Extra skill practice should be conducted.

c. **Efficient Performance.** The trainer must keep in mind that although movements are slowed, tasks take longer, and function checks, loading, unloading, and cleaning are affected by MOPP, it is a must that the gunner avoid damaging MOPP gear and risk possible exposure to lethal agents. Because of the great difference between no MOPP and MOPP4, gunners must be trained in all aspects of operation and maintenance of the weapon while practicing at the highest MOPP level. Only through repeated training and practice can the soldier be expected to perform all tasks efficiently.

CHAPTER 7

TRAIN THE TRAINER PROGRAM

The information in this chapter tells how to train the trainer, how to advise the trainer, and how to assist the trainer in preparing the crew for the three phases of MG marksmanship.

7-1. CONCEPT

The Train The Trainer Program must be planned properly and conclude with a certification program. In the planning phase, the instructor teaches the trainer to be flexible and thorough. He must plan alternate exercises in case weather or other constraints prohibit the originally scheduled training. Training must not follow rigid timetables that inhibit training and learning; instead, schedules should be established that provide sufficient time to correct mistakes and ensure learning. The following are factors that the trainers are taught to consider:

a. **Support.** Training requires support. The trainers must ensure that the training is conducted within the resource levels and that the training received justifies the material used. Public address systems should be used if the group is larger than a platoon-size element.

b. **Time.** Ample time must be allocated for each phase.

c. **Participants.** Trainers must consider whether or not the groups or individuals to be trained are capable of benefiting from the phases selected. (See paragraph 7-2a.)

d. **Safety.** Safety is the most important factor and therefore will be the main consideration. Trainers as well as the personnel to be trained will ensure all safety precautions are met. Anyone observing any unsafe act will immediately call CEASE FIRE or HALT to any training. A good way to emphasize safety is to give the students a test on the procedures. An example of a written examination on safety is shown in Figure 7-1, page 7-2.

SAFETY REQUIREMENT EXAM		
INSTRUCTIONS: Mark each answer as being true or false.		
T	F	
	√	l. Firing can be done without a safety officer.
	√	2. Firing can be done without correct headspace.
	√	3. Firing can be done without correct timing.
√		4. No gun should be loaded until the command to do so.
√		5. Each gun is cleared and checked after it is fired.
√		6.No one is allowed in front of the gun for any reason unless authorized by the OIC/NCOIC.
	√	7. Guns can be fired outside safety areas.
	√	8. Guns can be pointed at other personnel while clearing them.
√		9. Firing stops immediately upon the command CEASE FIRE.
√		10. Personnel will stay in designated safety areas until the command is given to leave by OIC/NCOIC.
Answer key: l. F; 2. F; 3. F; 4. T; 5. T; 6. T; 7. F; 8. F; 9. T; 10. T.		

Figure 7-1. Safety examination.

7-2. TRAINER CERTIFICATION PROGRAM

The Trainer Certification Program is designed to build pride, confidence, and overall working/teaching knowledge of the functions, employment, and overall training of the MG. The trainer must know how to do all the tasks he is going to be teaching; the certification program determines his capability to do this. The program is conducted in three phases.

a. **Phase I: Basic Fundamentals of Marksmanship.** In this phase the instructor must ensure the trainers being trained perform dry fire techniques; prepare firing positions, fighting positions, and range cards; and manipulate the T&E mechanism. At the end of this phase, they will be given a hands-on examination to test their basic knowledge. This test may be developed by the OIC/NCOIC of the MG committee.

b. **Phase II: Basic Marksmanship.** In this phase, the trainer moves one step closer to becoming a skilled gunner. He must learn the correct procedures to zero the weapon; he must master the 10-meter target paster; and once this is accomplished, he must be able to engage targets on a transition range. He must show his mastery in this phase by engaging targets at different ranges. Upon completion, the trainer is ready to move to the advanced stage of marksmanship.

c. **Phase III. Advanced Gunnery.** This is the final phase that the trainer has to achieve to standard.

(1) First, he must understand the procedures using the techniques of advanced gunnery.

(2) He must then learn tracking, which consists of maintaining correct alignment of the sights on a moving target by moving the gun at the same angular speed as that of the target.

(3) He will learn fire distribution next, which means he must be able to distribute fire over the entire target area.

(4) The last stage the trainer must learn is the correct assault fire techniques. During this stage, the gunner will normally fire on an FPL after the enemy has assaulted. He may also be given a PDF to fire at. Another assault method the trainer learns is to fire the weapon while it is mounted on a moving M113.

d. **Recertification Program.** All MG instructors are required to be recertified on a semi-annual basis. This will consist of the trainer being required to present a selected period of instruction within the current POI. The selection will be made by the CDR/OIC/NCOIC.

NOTE: Documentation of the results of the Trainer Recertification Program and the requirements must be maintained on file.

7-3. RESPONSIBILITIES AND DUTIES OF THE TRAINER

The trainer must be present during the planning and during any instruction given. The success of the preparation and instruction of all training depends upon the thoroughness with which the trainer performs the following duties:

- ŽHe must assist the gunner in targeting the MG.
- He should require the gunner to inspect his equipment and MG.
- He should explain and require the gunner to explain the exercise that he is about to perform.
- He should ensure the gunner's sight picture is accurate.
- He must observe the gunner's position, grip, and manipulation during any firing.
- He must show the gunner how to adjust his fire by observation.
- He must point out errors and explain their effect on the exercise.

NOTE: Safety is a must during the entire certification program. The OICs/NCOICs must ensure that no unsafe acts are tolerated.

* APPENDIX A

SAFETY

Safety precautions must be observed during all marksmanship training. This appendix recommends safety precautions for the ranges described in this manual. Range safety requirements will vary with the different courses of fire.

A-1. TRAINING RANGES

The following safety precautions will be observed when using any training range.

a. A red flag will be displayed at the entrance or in some other prominent location on the range during firing.

b. Firing limits will be marked with red and white striped poles visible to all firers.

c. Firing lanes must be clearly marked for easy identification.

d. All communication equipment, such as microphones, PA systems, loudspeakers, and radios, must be in good working condition.

e. When not in use, all MGs will be kept in a prescribed area with proper safeguards.

f. No smoking will be permitted near ammunition, explosives, or flammables. Firefighting equipment must be installed in ammunition shed or area.

g. Hearing protection devices will be worn by all personnel during firing.

h. Obstructions will never be placed in the muzzles of guns about to be fired.

i. Always assume that guns are loaded until they have been thoroughly examined and found to contain no ammunition.

A-2. RANGE PROCEDURES

Safety requirements must be followed by all personnel on the range before, during, and after firing. It is the responsibility of the safety officer, OIC, and safety NCO to ensure all procedures are met. They are as follows:

a. **Before Firing**.

(1) All prescribed roadblocks and barriers will be closed and necessary guards posted.

(2) All guns will be checked to ensure that they are clear of ammunition and obstructions, and that the covers are up to show they are cleared.

(3) All individuals will be briefed on the firing limits of the range and firing lanes.

(4) Range clearance will be obtained from the installation range-control office.

(5) The downrange area will be checked before firing to ensure that all personnel and equipment are clear of the area.

(6) The NCOIC or range safety officer places a complete first aid kit on the range.

(7) The NCOIC or range safety officer places medical personnel on the range or nearby to enable quick contact.

* (8) Due to the potential injury hazards associated with improper headspace and timing, either the NCOIC or RSO will assume personal responsibility for ensuring that all weapons are set for proper headspace and timing. He will do this before emplacing weapons on the firing line. This applies to all manually fired weapons. It does not apply to mechanically fired weapons.

*** WARNING**

Before placing any manually fired weapon on the firing line, personally check each one for proper headspace and timing.

(9) Soldiers handle guns only on the command of the tower operator or the OIC.

(10) Soldiers draw ammunition only on command of the OIC. When using two or more lots of ammunition, the OIC keeps the lots separate. He identifies ammunition by lot numbers in case of accident or malfunction. Soldiers should retain original packaging for possible repacking of ammunition.

(11) Protect all ammunition from the rays of the sun. Keep oil away from ammunition.

(12) Avoid moving forward of the firing line except with permission of the tower operator, range safety officer, or OIC.

b. **During Firing**.

(1) Should anyone note an unsafe condition, that person will immediately call CEASE FIRE. Firing resumes only when the OIC so directs.

(2) All personnel know the danger of moving forward of the firing line to score their targets. Before the firing line is cleared, and before anyone can go forward, the OIC or range safety officer clears all machine guns.

(3) In the event of a runaway gun, the firer should keep the barrel pointed downrange. He should leave the cover latched.

c. **During Darkness**.

(1) The NCOIC or range safety officer checks the downrange area before firing. He ensures that the area is clear of personnel and equipment. To make sure, he asks three times over a public address system, "Is anyone downrange?" pausing long enough each time to permit a response.

(2) A blinking red light must be used in addition to the red flag. It should be displayed at the entrance to the range or at some other prominent location.

(3) Two red lights will be mounted on the striped poles marking the limits of fire. They must be visible to all firers.

(4) No one will move from his position until told to do so by the OIC.

d. **After Firing**.

(1) Safety personnel will inspect all guns to ensure that they are clear. All guns on the firing line will be rodded whether they were fired or not. A check will be conducted to determine if any brass, links, or live ammunition is in possession of the troops.

(2) When guns have been cleared, they will be kept in a prescribed area with the bolt lock to the rear, and the feed cover raised.

A-3. WEAPONS MAINTENANCE

Before disassembly can be conducted, the gun must be cleared as discussed in Chapter 1.

a. Make sure the bolt is forward before removing the backplate assembly.

b. Never attempt to cock the gun while the backplate is off and the driving spring assembly is in place. If the backplate is off and the driving spring assembly is compressed, the retaining pin on the driving spring rod can slip from its seat in the sideplate and could cause serious injury to anyone behind the gun.

c. When lifting the receiver, barrel, or the tripod, hold it with a firm grip to avoid dropping it and possibly injuring someone.

APPENDIX B

TRAINING AIDS AND DEVICES

The use of devices in marksmanship training programs are important factors because they may allow the gunner to get an idea of what actual combat is like. The training devices designed for the MG are the M19 blank firing attachment, the multiple integrated liner engagement system, the sighting bar, and the M3 recoil amplifier barrel used with plastic ammunition.

B-1. BLANK FIRING ATTACHMENT

The M19 BFA was developed to permit the MG to fire the M1A1 blank cartridge in the automatic fire mode. The BFA is an easy to install, reliable device that allows the MG to be used more realistically during an FTX. The M19 BFA weighs 15.5 pounds and can be installed using either a crescent wrench or the blank/live round discriminator (a component of the BFA that has been designed to serve as a wrench). The design of the M19 will not allow a live round to be loaded while the blank/live round discriminator cover is in position. Normal headspace and timing must be made when firing blank rounds using the M19 (Figure B-1).

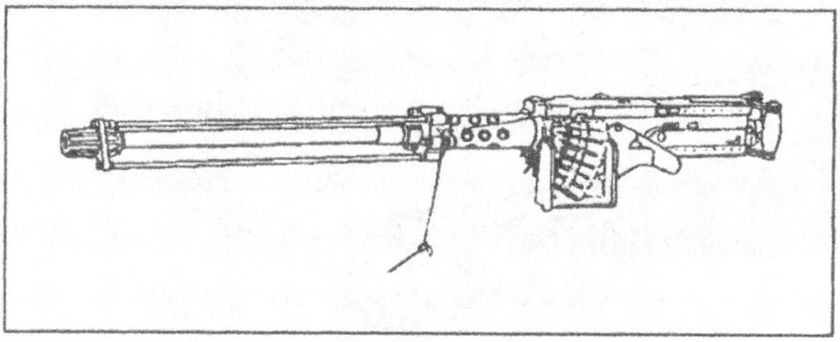

Figure B-1. M19 BFA mounted on MG.

WARNING
The flash of the M1A1 round fired with the M19 BFA at night is extremely bright. It could cause temporary night blindness, and night vision devices may be temporarily disabled.

B-2. MULTIPLE INTEGRATED LASER ENGAGEMENT SYSTEM

The MILES transmitter has been developed for the MG as well as for many other weapons. It consists of a lightweight laser transmitter that sends coded hit/kill messages whenever blank rounds are fired. The MILES transmitter is normally used as part of an M113 APC MILES kit; however, it can also be used on truck- or ground-mounted machine guns with modifications of the detector straps and the combat vehicle kill indicator (CVKI) light (Figure B-2).

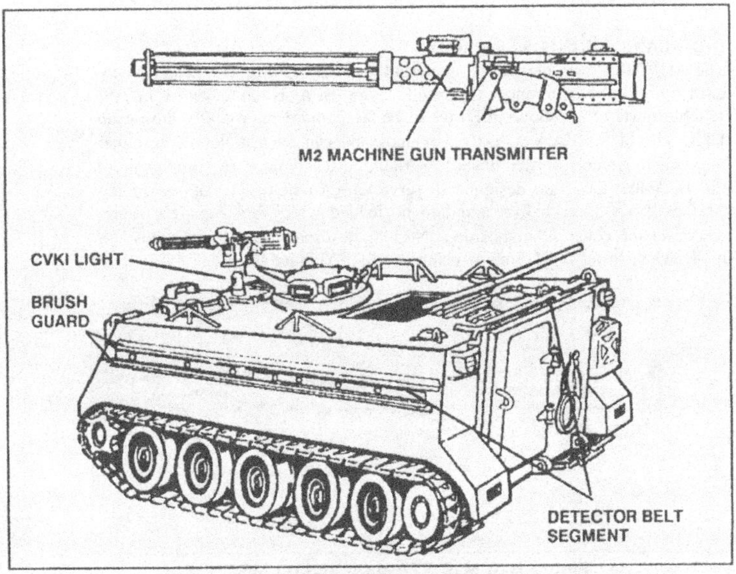

Figure B-2. Components of the M113 APC MILES kit.

a. Inspecting and Servicing the MILES Transmitter. Remove any dirt or oil from the lens using lens paper or a soft, dry cloth. Take care not to scratch the lens. Inspect the foam microphone cover; if it is wet or caked with dirt or blank firing residue, clean or replace it. Inspect the transmitter for indications of damage that would prevent normal operations (Figure B-3). Clean all surfaces. Ensure that the orange key is present. Use DA Form 2404 to report any damage to the transmitter.

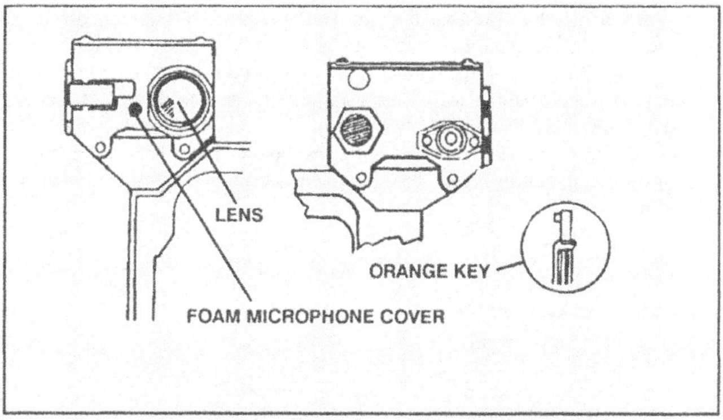

Figure B-3. Inspection of the MILES transmitter.

b. **Placing the Battery into the Transmitter.** Use the correct battery (NSN 6135-01-063-1978, BA 3090/U, 9V, alkaline); it will last about 100 hours. Flip open the latch; open the battery door and insert the battery; press the door closed; then press the latch closed (Figure B-4).

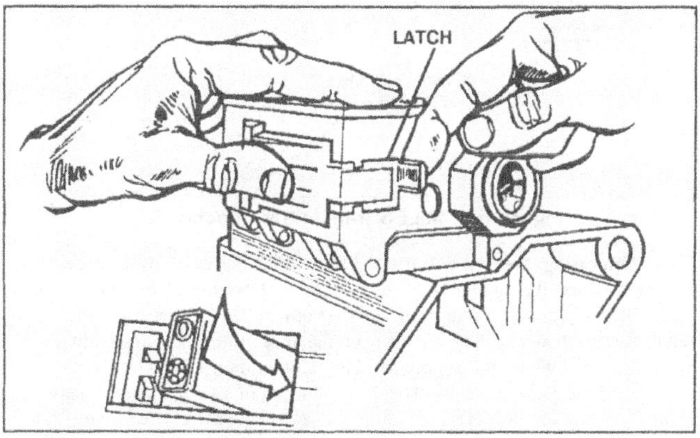

Figure B-4. Battery insertion.

c. **Attaching the Transmitter to the Machine Gun.** Attach the machine gun to its mount. Attach the M19 BFA to the machine gun. Unscrew the knob from the side of the transmitter bracket; swing the bottom plate down; and place the transmitter on the barrel support, flush against the receiver. The lip on the back of the mounting bracket must sit behind the barrel cooling bracket. Swing the bottom plate back up against the transmitter mounting bracket and tighten the knob securely with your hand (Figure B-5).

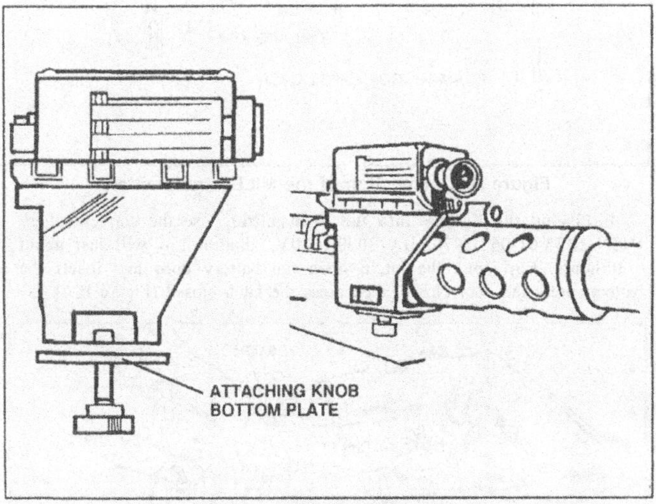

Figure B-5. MILES transmitter attachment.

d. **Operating the Transmitter.** Ensure that the M19 BFA is correctly mounted. Turn the orange key to WEAPON ON. Load the machine gun with M1A1 blank ammunition only (Figure B-6). Fire normally. The sound of the blanks firing will trigger the transmitter. The transmitter will work only as long as the supply of blank ammunition lasts. Once a day, use a clean cloth to remove blank fire residue from the transmitter lens. To test the transmitter, fire a shot burst and watch the firing lamp (Figure B-7). If no light, replace the battery and test again. If still no light, replace the transmitter. Use DA Form 2404 to report the information.

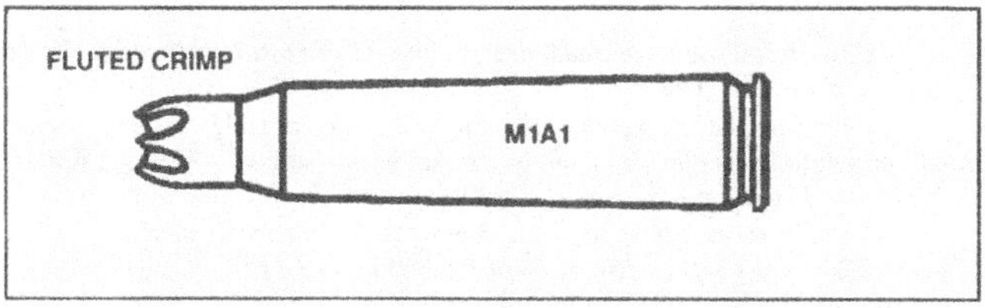

Figure B-6. Proper blank ammunition to use with M19 BFA.

> **WARNING**
> **Do not use ball ammunition or M1 blank ammunition with the BFA. Use only M1A1 blanks with the fluted, crimped case mouth.**

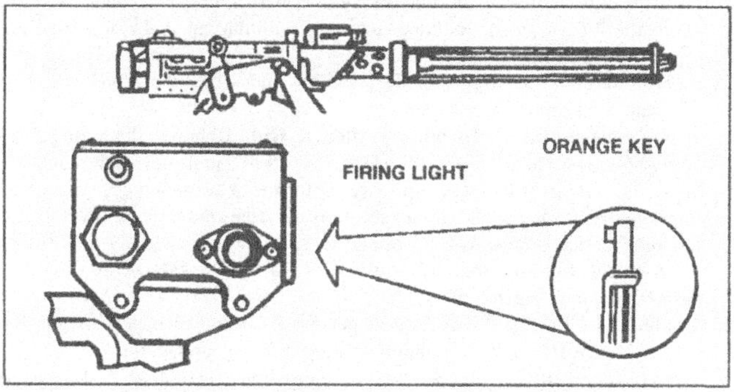

Figure B-7. Operation of the MILES transmitter.

e. **Aligning the Transmitter to the Machine Gun.** The transmitter must be aligned in the dry fire mode, using the dry fire trigger cable. To align the transmitter, you need a soldier wearing a MILES helmet and torso harness. The gun should already be mounted on a tripod and have the T&E attached.

(1) Position the soldier 50 to 75 meters away from the machine gun. Have him place a green controller key in his MILES receptacle and turn the key to the CONTROLLER position.

(2) Connect the trigger cable assembly to the receptacle on the rear of the MILES transmitter. Use the controller key to reset the system.

(3) Insert an orange weapon key into the transmitter receptacle and turn to WEAPON ON.

(4) Fire at the soldier by pressing the trigger cable button. Traverse the gun left and right, up and down, until the soldier's buzzer signals a "near miss." Hold the gun in that position.

(5) While he is being fired on, have the soldier move to his right until his buzzer stops. Have him mark that location.

(6) Have the soldier move to his left until the buzzer stops again. Have him mark that location.

(7) Have the soldier estimate the center between the two marks, and stand there.

(8) Cease fire. Without moving the barrel of the machine gun, adjust the sights until the soldier is seen through the rear sight aperture.

(9) Begin firing again and move the barrel up and down, noticing where the buzzer stops. Estimate the center of this up and down area and adjust the sight elevation to that point.

(10) Remove the trigger cable assembly and screw the protective cover onto the transmitter receptacle.

f. **Resetting the Transmitter After a Kill.** Remove the orange key from the transmitter. Use it to silence the "kill" indicator and the CVKI light. The transmitter will not fire with the key removed. When the controller resets the "kill" indicator, turn the transmitter to WEAPON ON and reinsert the orange key. Continue the mission.

g. **Using Precautions.** The following safety considerations apply to the M1A1 blank cartridge:

(1) Never fire blanks directly at personnel within 20 meters of the gun.

(2) Never fire blanks without wearing hearing protection.

(3) Never fire blanks when the temperature is below 0 degrees or above 120 degrees Fahrenheit.

(4) Never stand directly to the side of the weapon muzzle when blanks are fired.

(6) Never fire blanks in bursts in excess of 20 rounds.

(7) Always allow a cooling period between bursts to avoid overheating the gun.

B-3. SIGHTING BAR

The sighting bar is a locally fabricated device used for practicing the sighting and aiming exercise (Figure B-8). Paragraph 5-9 explains its use. This exercise requires the soldiers to look through the peephole and center on the front sight blade. The gunner adjusts the peephole until he is able to align the front sight blade in the center of it. The instructor will then review and critique the gunner on sight alignment. The device is also equipped with a scaled target for practicing sight picture. Once sight alignment is mastered, the gunner will be required to show a correct sight picture. This is done by adjusting the scaled target until the front sight plate is bottom center of it. Again the instructor will review and critique the gunner. This exercise will continue until the gunner has mastered the technique of sighting and aiming; this requires a lot of practice.

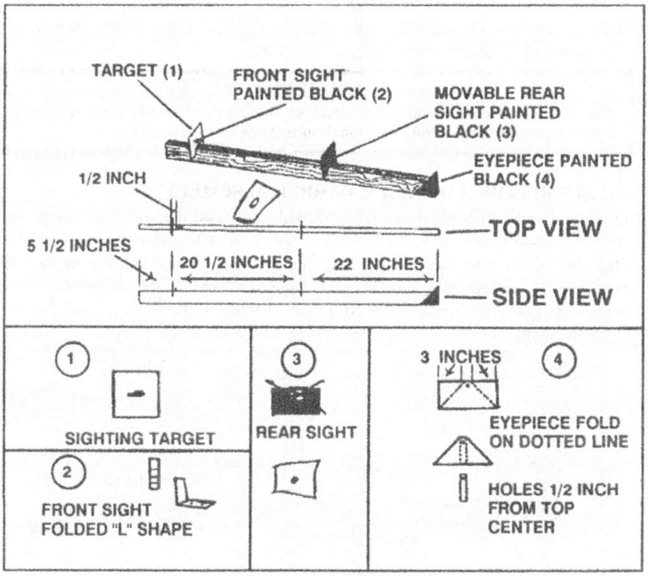

Figure B-8. Sighting bar.

B-4. SHORT RANGE TRAINING AMMUNITION

To augment the use of the caliber .50 ammunition, a new generation of SRTA (the M858 ball and the M860 tracer) has been developed.

a. The caliber .50 ball and tracer cartridges have configurations and dimensions similar to those of the corresponding service rounds; however, the base of the cartridge, which encloses the primer, is made of aluminum. The rest of the case is-made of a blue-colored plastic material. The tip of the tracer projectile is colored red. The complete round weighs about 1.05 ounces and is 5.19 inches long, which is slightly shorter than the service round. The cartridges are linked with M9-type metallic links for use with the MG. Muzzle velocity of the projectile is about 4,000 feet per second. The light weight of the projectile, about 50 grains, and its low sectional density cause its velocity to drop rapidly.

b. The SRTA allows training in small local training areas without fixed training facilities, in MOUT facilities, and in combat training theaters. With its 150-meter usable range, the SRTA can be used on grouping, zeroing, and 10-meter scaled silhouette firing.

WARNING

The SRTA projectile causes considerable damage out to 150 meters and is considered dangerous out to its maximum range of 700 meters.

B-5. M3 RECOIL AMPLIFIER BARREL ASSEMBLY

An M3 RABA (Figure B-9) is used with the machine gun for firing the training ammunition. It is assembled to the gun in the same manner as the standard barrel. The RABA provides the means to boost the power that is required to recoil the barrel and effectively cycle the weapon when relatively underpowered M858 SRTA or M860 SRTA-T are fired in the machine gun.

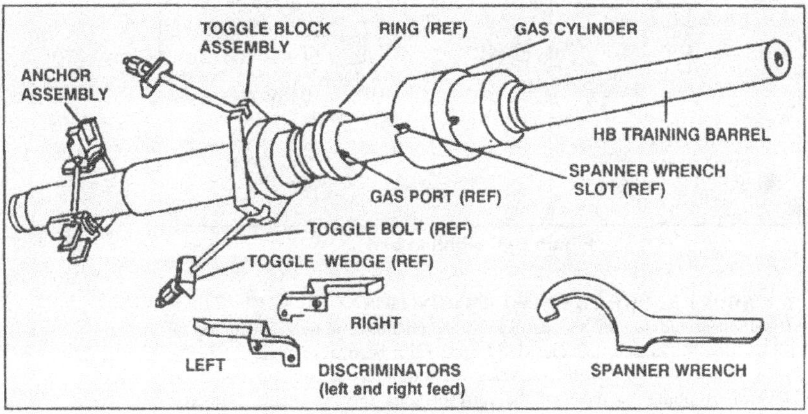

Figure B-9. M3 RABA.

APPENDIX C

RANGES AND TRAINING EXERCISES

This appendix provides guidance for US Army machine gun exercises. It is designed to be a working guide for trainers and commanders by explaining range layouts, conduct of firing exercises, and qualification standards. The two basic firing ranges will be the multipurpose MG transition range and the multipurpose range complex.

* Section I
MULTIPURPOSE MACHINE GUN TRANSITION RANGE

All machine gun firing can be conducted on the multipurpose MG transition range. It allows for complete firing of the 10-meter course as well as transition, day, night, and NBC firing. The firing area has 10 lanes; its layout is shown in Figure C-1. Details on the setup and target configuration are described in FM 25-7. The minimum personnel required to operate the range are as follows: OIC, NCOIC, safety officer, ammunition NCO, tower operator, lane NCOs, primary instructor, and concurrent training instructors. Local policy may require more personnel.

* The targets listed in this section are only recommendations. Units may alter the types of targets and ranges used depending on availability.

- *For night qualification,* units with the AN/PEQ-2A, ANPAS 13 or the AN/TVS-5 upgraded with the third-generation tube should use Appendix G tables and qualification standards.
- *For day qualification,* leaders will use this section until the STRAC (AR 350-38) changes, at which time they will use only Appendix G.

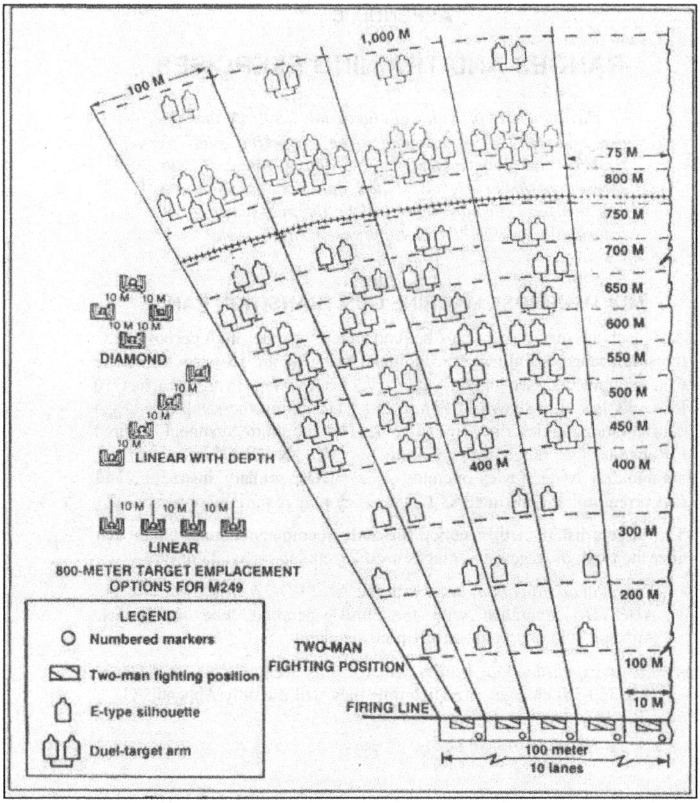

Figure C-1. Multipurpose machine gun transition range layout.

C-1. TEN-METER RANGE LAYOUT

The standard, basic 10-meter range can accommodate a unit of 200 to 250 soldiers at a time; however, concurrent training may be required. This range (Figure C-2) can be used to zero the M249, M60, and M2 machine guns, and to fire the 10-meter portion of qualification. It is also used to familiarize soldiers with the characteristics, noise, and recoil of the weapon. This is the range used to practice target observation and adjustment of fire, to practice MG traversing and searching, to develop speed during operation, and to obtain an accurate burst.

a. **Characteristics.** The following data applies to the standard 10-meter range.

Number of firing positions	**20 lanes.**
Firing lane width	**60 meters (3 meters per lane).**
Target area width	**60 meters at the farthest targets (10 meters).**
Firing point configuration	**Numbered markers on slightly elevated and sodded ground; brass deflectors between lanes.**
Target configuration	**Standard 10-meter machine gun target stretched over a wooden frame, one on each lane.**
Associated facilities	**Standard facilities, public address system, and bleachers.**

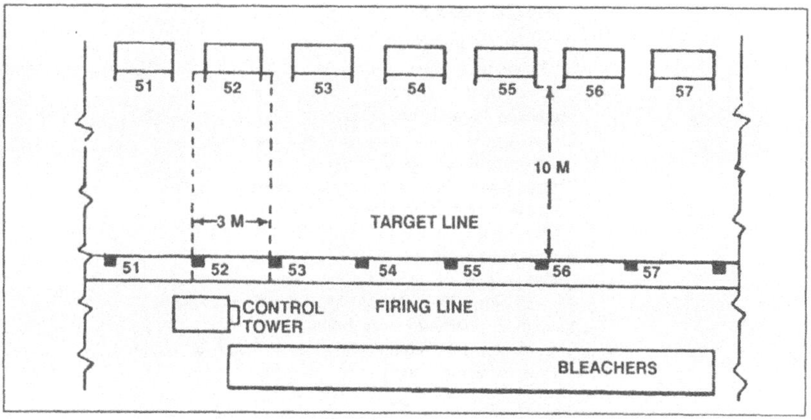

Figure C-2. Ten-meter range layout.

b. **Setup of Range.** The 10-meter range should meet the following requirements:

(1) The firing line should be long enough to emplace 20 MGs, allowing 3 meters between positions. For control, each position should be numbered.

(2) The target line should be 10 meters in front of the firing line. The MG marksmanship target is a paper target pasted onto target cloth that is stretched over a wooden frame. One target is set up for each position and is numbered to correspond with the numbered position.

(3) The instruction site for this range is the bleachers, which should be to the rear of the firing line.

(4) The control tower is located to the immediate rear and center of the firing line.

c. **Personnel Required.** Each range must be staffed with the following personnel:

- One officer in charge.
- One safety officer.
- One principal instructor.
- One assistant instructor for every 10 students.
- One ammunition NCO.
- One tower operator.
- Medical personnel.

d. **Equipment Required.** Although more equipment may be required by local range regulations, safety regulations, or unit SOPs, the minimum equipment required to operate the range is as follows:

- One public address system.
- One MG for each firing lane and one extra gun for every five lanes as a backup.
- One cleaning rod per assistant instructor.
- One scorecard per soldier.
- One asbestos glove for every two weapons.
- One M3 tripod for each lane.
- One caliber .50 pintle for each lane.
- One caliber .50 T&E mechanism for each lane.
- Medical evacuation capability.
- Communication equipment (wire or radio as required to operate the range).

* C-2. TEN-METER FIRING EXERCISES

The purpose of 10-meter firing is to develop skills in the delivery of initial bursts on target. When conducted properly, it will train gunners in the basic skills of zeroing, controlling the burst, traversing, and traversing and searching techniques. This paragraph discusses firing the 10-meter exercise on the multipurpose transition range; however, it is conducted the same on a basic 10-meter range except that the groups will be broken down to fit the firing lanes. The course is designed to fire the tripod-mounted M2. The basic MG target is used for all 10-meter firing and it allows four gunners to use it for practice or two gunners for practice and qualification. Those units wanting to fire qualification right after practice should use section A for practice and B for qualification for the first gunner, and sections C and D for the second gunner. Ten-meter firing will be conducted IAW Firing Table I (Figure C-3).

TASK	CONDITIONS TARGET/SITUATION	AMMO	STANDARD_____
1. Zero the M2 machine gun at 10 meters.	Instruct the gunner to engage scoring paster 1 in his assigned section with 6 single rounds in two 3-round shot groups and then paster 2 with the remaining rounds.	12 single rounds of 12.7-mm	Gunner must impact 4 of 6 rounds of the 3-round shot groups within paster 2 of his assigned section.______
2. Engage a 10-meter target using controlled-burst techniques.	Instruct the gunner to engage paster 3 in his assigned section using controlled bursts, then engage paster 4 in the same manner. Gunner will use 5- to 7-round bursts for this engagement.	two 7-round belts of 12.7-mm	Gunner must impact one controlled burst on pasters 3 and 4 in his assigned section. (No score required.)____
3. Engage scoring pasters 5 through 6 using the traverse and search technique of fire.	Instruct the gunner to engage pasters 5 through 6 using traverse and search technique of fire. Gunner will use 5- to 7-round bursts for this engagement.	35-round belt of 12.7-mm	Gunner must impact one round on each scoring paster of 5 through 6 in his assigned section. (No score required.)____
4. Engage scoring pasters 7 through 8 using the traverse and search technique of fire.	Instruct the gunner to engage pasters 7 through 8 using traverse and search technique of fire. Gunner will use 5- to 7-round bursts for this engagement.	56-round belt of 12.7-mm	Gunner must impact one round on each scoring paster of 7 through 8 in his assigned section ______. (No score required.)____
5. Engage linear target at 10 meters using traverse fire technique.	Instruct the gunner to engage pasters 1 through 4 in his assigned section using traverse fire. Gunner will use 5- to 7-round bursts for this engagement.	28-round belt of 12.7-mm	Gunner must impact at least four rounds on each scoring paster 1 through 4 in his assigned section ______.
6. Engage scoring pasters 7 through 8 using the traverse and search technique of fire.	Instruct the gunner to engage pasters 7 through 8 using traverse and search technique of fire. Gunner will use 5- to 7-round bursts for this engagement.	56-round belt of 12.7-mm	Gunner must impact at least four rounds on each scoring paster 7 through 8 in his assigned section ______.
7. Engage scoring pasters 5 through 6 using the traverse and search technique of fire.	Instruct the gunner to engage pasters 5 through 6 using traverse and search technique of fire. Gunner will use 5- to 7-round bursts for this engagement.	35-round belt of 12.7-mm	Gunner must impact at least four rounds on each scoring paster 5 through 6 in his assigned section ______.

*** Figure C-3. Ten-meter firing table (Firing Table I).**

a. **Objective**. In order for the M2 gunner to become proficient with the machine gun, the following objectives must be considered:

- Obtain an accurate initial burst.
- Master the technique of controlled burst firing.
- Traverse and search the machine gun effectively.
- Observe and adjust fire.
- Operate with speed.

b. **Organization**. The unit is organized into groups of 10. Each group is assigned a firing order number. One order becomes gunners and the other assistant gunners. The gunners and assistant gunners are assigned lanes and required to set up their guns and perform pre-fire checks.

(1) Ten-meter practice fire (Tasks 1 through 4).

(a) The gunner is required to set his rear sight at 500 yards. On command, the first order zeros using scoring pasters 1 and 2 of their assigned section.

(b) At the completion of zeroing, the gunners are required to engage the next two aiming pasters (scoring pasters 3 and 4 of the same section), using controlled bursts, and then area targets 5 through 6 and 7 through 8, using traverse and search techniques. Time is allowed between bursts to permit gunners to observe and adjust their fire.

(c) After the first order has fired, the second order fires the course in the same manner using the next assigned section. After both orders have fired, the guns are cleared.

(2) Ten-meter qualification (Tasks 5 through 7).

(a) The gunner is then required to set his rear sight at 550 yards. On command, the first order fires at scoring pasters 1 through 4 of their assigned section.

(b) At the completion of their linear engagement, commands are given to require the gunners to engage area targets 5 through 6 and then 7 through 8, using traverse and search techniques. Time is allowed between bursts to permit gunners to observe and adjust their fire.

(c) After the first order has fired qualification, the second order fires the course in the same manner using the next assigned section. After both orders have fired, the guns are cleared.

(2) *Ten-meter qualification (Tasks 5 through 7).*

(a) The gunner is then required to set his rear sight at 550 yards. On command, the first order fires at scoring pasters 1 through 4 of their assigned section.

(b) At the completion of their linear engagement, commands are given to require the gunners to engage area targets 5 through 6 and then 7 through 8, using traverse and search techniques. Time is allowed between bursts to permit gunners to observe and adjust their fire.

(c) After the first order has fired qualification, the second order fires the course in the same manner using the next assigned section. After both orders have fired, the guns are cleared.

c. **Target Analysis.** The exercises use the basic machine gun target (FSN 6920-078-5128) (Figure C-4). The following explanation of the target, including the size of the aiming pasters and scoring spaces, will aid in zeroing the M2 and will facilitate control during firing exercises.

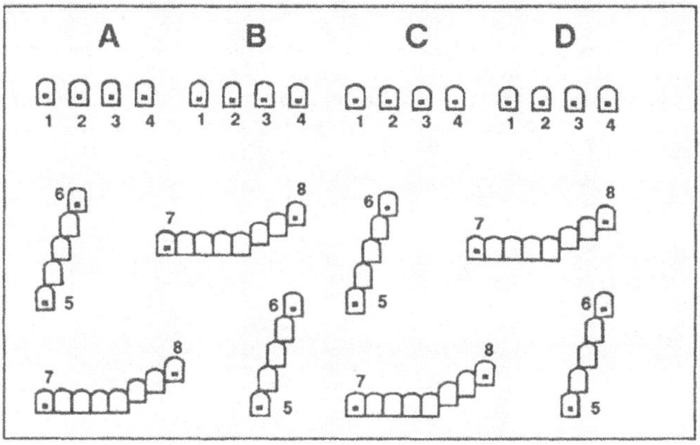

Figure C-4. Basic machine gun target.

(1) The target consists of four sections lettered A, B, C, and D. Each section has four point targets numbered 1, 2, 3, and 4; and two sets of area targets numbered 5 through 6 and 7 through 8. Each space is 4 centimeters wide and 5 centimeters high. The black aiming pasters within some numbered scoring spaces are 1-centimeter squares. Firing at targets 1 through 4 allows the gunner to use the traverse-fire technique

targets 5 through 6 and 7 through 8 give them practice in traverse and search techniques.

(a) Point targets. Point targets on the basic machine gun target are considered to be pasters 1 through 4 of section A, B, C, and D. Firing at point targets exposes the gunner to zeroing techniques and controlled-burst fire techniques. Targets 1 through 4 will also be used as a linear target for qualification.

(b) Area targets. Area targets on the basic machine gun target are considered to be pasters 5 through 6 and 7 through 8 of section A, B, C, and D. Target group 5 through 6 exposes the gunner to T&E manipulation when using the tripod mount and body position changes to engage targets in depth (elbow position changes). This causes him to use a series of aiming points to disburse fire across the target when using the tripod. Target group 7 through 8 exposes the gunner to position changes to engage linear targets with depth. It causes him to control the burst length from the weapon, use a series of aiming points, and disburse fire across the target.

(2) Targets are analyzed and scored to determine the gunner's proficiency and to see if more training is needed in any of the fundamentals of M2 gunnery. During firing with a properly zeroed weapon, a target is best analyzed by considering the common errors of M2 gunnery (Figure C-5).

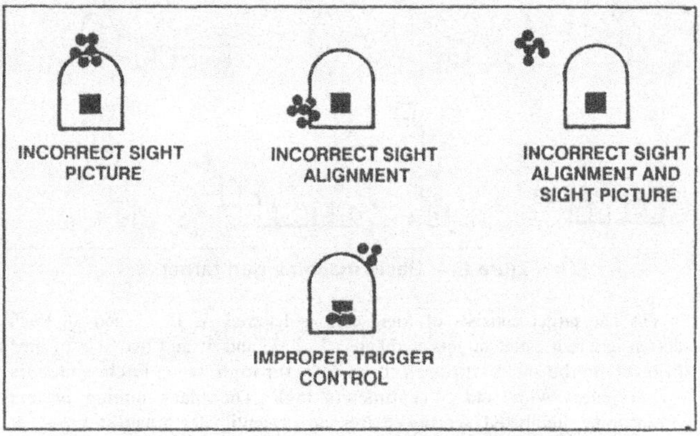

Figure C-5. Common errors found on the basic MG target.

NOTE: Large shot groups are usually caused by incorrect position and grip; small shot groups outside of the scoring space are usually caused by incorrect sight alignment, sight picture, or zero.

d. **The 10-Meter Zero.** Ten-meter zeroing is accomplished by adjusting the rear sight until the strike of the projectile coincides with the point of aim at a given range. On the 10-meter range, the elevation scale must be set at 500 yards so that it can impact on the desired aiming point. Before zeroing the weapon, the gunner must first center the rear sight for windage by aligning the index with the center of the windage scale.

e. **Controlled Burst Firing.** This firing exercise reinforces the dry-fire experience and allows the gunner practice in firing of controlled bursts and in providing the most accurate and tightest shot groups obtainable with the M2.

f. **Ammunition Issue.** The ammunition for the firing of this exercise will be broken down at the ammunition point and issued to each assistant gunner as he is assigned a firing point. The total ammunition for this exercise is 236 rounds (117 for practice and 119 for qualification). Issue will consist of the following rounds or belts:

- Task 1, 12 single rounds
- Task 2, two 7-round belts.
- Task 3, one 35-round belt.
- Task 4, one 56-round belt.
- Task 5, one 28-round belt.
- Task 6, one 56-round belt.
- Task 7, one 35-round belt.

g. **Firing Sequence.** Firing of Table I will be conducted in the following manner. These procedures pertain to firing with a single gunner using sections A and B; if there is a second gunner, he will use C and D.

(1) *Task 1, Zero.*

(a) The gunner will fire 3 rounds, single shot, at paster A1.

(b) The gunner will then move down range to observe the shot group and triangulate it. No adjustments to the weapon will be made at this time.

(c) The gunner will then fire another 3 single rounds at A1.

(d) The gunner then goes down range to observe the shot group, triangulate it, and make necessary adjustments to his weapon.

(e) The gunner repeats steps (c) and (d), but shoots at paster A2.

NOTE: If the gunner is able to zero his weapon using 9 rounds, have him use the remaining 3 to confirm his zero. If the gunner is unable to zero within 12 rounds, remove him from the firing line and give him remedial training.

(2) *Task 2, Controlled Burst Firing.*

(a) The gunner will fire the first burst of 5 to 7 rounds at paster 3.

(b) The gunner then goes down range to observe, marks the projectile holes, and analyzes his burst.

(c) The gunner repeats steps (a) and (b), but fires at paster 4 of same section.

NOTE: The primary purpose of this task is to get the gunner to fire 5- to 7-round bursts. Impact on the target is desired, but not necessary to complete this task. The gunner will fire a 5- to 7-round burst at each paster.

(3) *Task* 3, *Traverse and Search Fire.*

(a) The gunner will engage pasters 5 through 6, section A, firing a 5- to 7-round burst for each paster, using the traverse and search technique.

(b) The gunner will then move down range to observe and analyze his targets.

NOTE: These tasks are fired from the tripod firing position. If the gunner shows difficulty in manipulating the weapon, remove him from the firing line and give him remedial training.

(4) *Task 4, Traverse and Search Fire.*

(a) The gunner will engage pasters 7 through 8, section A, firing a 5- to 7-round burst at each paster, using the traverse and search technique.

(b) The gunner will then move down range to observe and analyze his targets.

(5) *Task 5, Traverse Fire.*

(a) The gunner will engage pasters 1 through 4, section B, firing a 5- to 7-round burst at each paster, using the traverse technique.

(b) The gunner will then move down range to observe and analyze his targets.

(6) *Task 6, Traverse and Search Fire.*

(a) The gunner will engage pasters 7 through 8, section B, firing a 5- to 7-round burst at each paster, using the traverse and search technique.

(b) The gunner will then move down range to observe and analyze his targets.

(7) *Task 7, Traverse and Search Fire.*

(a) The gunner will engage pasters 5 through 6, section B, firing a 5- to 7-round burst at each paster, using the traverse and search technique.

(b) The gunner will then move down range to observe and analyze his targets.

DANGER
Before any gunner goes down range, the lane NCO must ensure that the weapon has been properly cleared. Anyone observing any unsafe acts should immediately call a cease fire and notify range personnel.

h. **Scoring Procedures.** When scoring the 10-meter target, all scoring spaces are scored (1 through 4, 5 through 6, and 7 through 8). One point is given for each round impacting within each space. Rounds touching the boundary of a scoring space are considered hits, but they can be counted in only one scoring space. When firing at 1 through 4 with 28 rounds, the maximum score is 28 points. In group 5 through 6, five scoring spaces are engaged with 35 rounds with a possible score of 35 points. When firing groups 7 through 8, eight scoring spaces are engaged with 56 rounds with a possible score of 56 points.

(1) The total possible score of both groups is 119 points. A minimum of 84 points is required to pass the 10-meter course of fire.

(2) Soldiers failing to achieve minimum standards must be retrained and retested in a dry-fire mode until proficiency is demonstrated. They may not move on to transition firing. Soldiers should then re-fire the 10-meter portion with close supervision and coaching to ensure that the fundamentals are applied properly during live fire.

C-3. TRANSITION FIRING EXERCISES

Transition firing of the M2 machine gun will teach the gunner some techniques of fire that he may encounter in combat situations. The gunner will field zero his weapon and also incorporate the techniques of fire during limited visibility and NBC environments. The gunner needs to be aware that during certain situations his capabilities will be degraded. He is expected to compensate for these situations by applying all available techniques. Within this training, the gunner will be required to apply all the fundamentals of gunnery learned in preparatory gunnery training and 10-meter firing. Instructors should encourage gunners to perform immediate action if a stoppage occurs during fire. This procedure may be modified if local policies require the gunners to notify the range personnel first.

a. **Objectives.** The objectives of transition fire are —

- To engage targets at long ranges with the tripod-mounted machine gun.
- To understand the characteristics of fire.

Ž To field zero the M2 machine gun.

Ž To use range estimation to determine the distance to targets.

Ž To apply the method of adjusted point of aim.

b. **Organization.** Organization of a unit for transition firing is conducted the same as for 10-meter firing. Field zeroing is the first firing task of transition day-fire phase. The transition day-fire portion of the course must be fired to become a qualified M2 gunner.

c. **Ammunition Issue.** The ammunition for the firing of this exercise will be broken down at the ammunition point and issued to each assistant gunner as he is assigned a firing point. The total ammunition for this exercise is 182 rounds, and issue will consist of two belts; one 28-round belt and one 154-round belt.

d. **Firing Sequence.** Firing of Table II (Figure C-6) will be conducted in the following manner. These procedures pertain to firing with a single gunner using a tripod-mounted M2. If there is a second gunner, he will use the same procedures.

(1) *Task 1, Field Zero.*

(a) The gunner must first center the rear sight in the same manner as sight setting for 10-meter firing. He selects the appropriate range mark with his elevation knob.

(b) The gunner will load one 28-round belt of ammunition and fire a burst of 5 to 7 rounds at the 550-meter, double E-type silhouette.

(c) If the impact of the rounds (beaten zone) is over the target, the gunner has probably fired with an incorrect sight picture (too high on the target). He needs to relay the gun on the original point of aim and then he relays back on the target and fires again. If the gunner again fails to zero with a proper sight picture, he should have the unit armorer inspect the weapon before continuing the zeroing procedure.

(d) The gunner will observe the beaten zone and make adjustments to the sights so the round will impact on the target.

(e) After adjustments are made, the gunner will then relay on the center base of the target and fire another 5- to 7-round burst.

(f) The gunner will repeat steps c, d, and e with the remaining rounds.

NOTE: If the gunner is unable to zero within 28 rounds, he is removed from the firing line and given remedial training.

(2) *Task 2 through Task 8, Single and Multiple Engagements.*

(a) The gunner will load one 154-round belt of ammunition and fire bursts of 5 to 7 rounds at double E-type silhouettes at ranges of 800, 400, 700, and 1,000 meters at vehicular targets. He will also engage multiple double E-type silhouettes targets at 400 to 700; 550 and 800; and 400, 550, and 1,000 meters respectively.

(b) The gunner will observe the beaten zone and make adjustments as necessary to hit the targets.

(c) After firing is completed, weapons will be cleared and the gunner critiqued. If there is a second gunner to fire, the entire procedure will be repeated.

NOTE: After both gunners have fired for practice, they will then be assigned new lanes and issued more ammunition for record firing. If possible, the gunner should keep the same weapon he used for practice fire. The only change in the above procedures will be in step (c); here, the gunner will be scored along with the critique. If the gunner is unable to meet standard, he is removed from the firing line, given remedial training, and re-fired.

DANGER

Before any gunner changes lanes, the lane NCO must ensure that the weapon has been properly cleared. Anyone observing any unsafe acts should immediately call a cease fire and notify range personnel.

TASK	CONDITIONS TARGET/SITUATION	AMMO	STANDARD
1. Zero the M2 machine gun.	Gunner is in stationary tripod firing position and engages a 550-meter, double E-type silhouette. Gunner will use 5-to 7-round bursts for this engagement.	28-round belt of 12.7-mm.	Gunner must impact one burst on the 550-meter target.
2. Engage a single, double E-type silhouette.	Gunner is in stationary tripod firing position and engages an 800-meter, double E-type silhouette. Gunner will use 5-to 7-round bursts for this engagement.	14 rounds of 12.7-mm.	Gunner must impact one burst on the 800-meter target within 20 seconds.

Figure C-6. Practice/Qualification Table II.

TASK	CONDITIONS TARGET/SITUATION	AMMO	STANDARD
3. Engage a single. double E-type silhouette.	Gunner is in stationary tripod firing position and engages a 400-meter, double E-type silhouette. Gunner will use 5-to 7-round bursts for this engagement.	14 rounds of 12.7-mm	Gunner must impact one burst on the 400-meter target within 20 seconds.
4. Engage a single. double E-type silhouette.	Gunner is in stationary tripod firing position and engages a 700-meter, double E-type silhouette. Gunner will use 5-to 7-round bursts for this engagement.	14 rounds of 12.7-mm	Gunner must impact one burst on the 700-meter target within 25 seconds.
5. Engage a single. double E-type silhouette.	Gunner is in stationary tripod firing position and engages a 1,000-meter, vehicle type silhouette. Gunner will use 5-to 7-round bursts for this engagement.	14 rounds of 12.7-mm	Gunner must impact one burst on the 1,000-meter target within 25 seconds.
6. Engage multiple double E-type silhouettes.	Gunner is in stationary tripod firing position and engages 400-and 700-meter, double E-type silhouettes. Gunner will use 5-to 7-round bursts for this engagement.	28 rounds of 12.7-mm.	Gunner must impact one burst on each target within 35 seconds.
7. Engage multiple double E-type silhouettes.	Gunner is in stationary tripod firing position and engages 550-and 800-meter, double E-type silhouettes. Gunner will use 5-to 7-round bursts for this engagement.	28 rounds of 12.7-mm.	Gunner must impact one burst on each target within 35 seconds.
8. Engage multiple double E-type silhouettes.	Gunner is in stationary tripod firing position and engages 400-, 550-and 1,000 meter, double E-type and vehicle silhouettes. Gunner will use 5-to 7-round bursts for this engagement.	42 rounds of 12.7-mm.	Gunner must impact one burst on each target within 45 seconds.

Figure C-6. Practice/Qualification Table II, continued.

e. **Scoring Procedures.** One point is given for each target hit with an additional 2-point bonus if the target is hit on the first burst. The total possible points for day fire is 33. A minimum 7 points (7 out of 11 exposures) is required to meet standard on the day-fire course.

C-4. NBC FIRING

Since NBC plays an important part in our preparation for war on the modern battlefield, it is important that each soldier is prepared to accomplish the mission even if the area is contaminated and he must wear protective gear.

a. **Objectives.** The objectives of this training are:

- To teach the gunners to take appropriate action when notified of a contaminated area.
- To identify targets while in a contaminated area.
- To engage targets while in appropriate NBC gear.

b. **Organization.** The unit is organized the same as in paragraph C-2b, except both the gunner and assistant gunner must wear protective masks. Protective clothing is optional.

c. **Ammunition Issue.** The ammunition for the firing of this exercise will be broken down at the ammunition point and issued to each assistant gunner as he is assigned a firing point. The total ammunition for this exercise is 182 rounds, and issue will consist of two belts; one 28-round belt and one 154-round belt.

d. **Firing Sequence.** Firing of Table III (Figure C-7, pages C-16 and C-17) will be conducted in the following manner. These procedures pertain to firing with a single gunner using a tripod-mounted M2. If there is a second gunner, he will use the same procedures.

(1) *Task 1, Field Zero.*

(a) The gunner must first center the rear sight in the same manner as setting the sights for 10-meter firing. He selects the appropriate range mark with his elevation knob.

(b) The gunner will load one 28-round belt of ammunition and fire a burst of 5 to 7 rounds at the 550-meter, double E-type silhouette.

(c) If the impact of the rounds (beaten zone) is over the target, the gunner has probably fired with an incorrect sight picture (too high on the target). He should re-fire concentrating on the proper point of aim and sight picture. If the impact is still off, have the unit armorer correct the calibration and the gunner repeat the zeroing procedure.

(d) The gunner will observe the beaten zone and make adjustments to the sights so the round will impact on the target.

(e) After adjustments are made, the gunner will then relay on the center base of the target and fire another 5- to 7-round burst at the same target.

(f) The gunner will repeat steps c, d, and e with the remaining rounds.

NOTE: If the gunner is unable to zero within 28 rounds, he is removed from the firing line and given remedial training.

(2) *Task 2 through Task 8, Single and Multiple Engagements.*

(a) The gunner will load one 154-round belt of ammunition and fire bursts of 5 to 7 rounds at double E-type silhouettes at ranges of 800,400,

700, and a 1,000-meter vehicular target. He will also engage multiple, double E-type silhouettes targets at 400 and 700; 550 and 800; and 400, 500, and 1,000 meters, respectively.

(b) The gunner will observe the beaten zone and make adjustments as necessary to hit the targets.

(c) After firing is completed, weapons will be cleared and the gunner critiqued. If there is a second gunner to fire, the entire procedure will be repeated.

NOTE: After both gunners have fired for practice, they will then be assigned new lanes and issued more ammunition for record firing. If possible, the gunner should keep the same weapon he used for practice fire. The only change in the above procedures will be in the last step; here, the gunner will be scored along with the critique. If the gunner is unable to meet standard, he is removed from the firing line, given remedial training, and re-fired.

DANGER

Before any gunner changes lanes, the lane NCO must ensure that the weapon has been properly cleared. Anyone observing any unsafe acts should immediately call a cease fire and notify range personnel.

e. **Scoring Procedures.** One point is given for each target hit with an additional 2-point bonus if the target is hit on the first burst. The total possible points for day fire is 33. A minimum of 7 points (7 out of 11 exposures) is required to meet standard on the NBC fire course.

TASK	CONDITIONS TARGET/SITUATION	AMMO	STANDARD
1. Zero the M2 machine gun.	Gunner is in stationary tripod firing position and engages a 550-meter, double E-type silhouette. Gunner will use 5-to 7-round bursts for this engagement.	28-round belt of 12.7-mm.	Gunner must impact one burst on the scaled, 550-meter target.
2. Engage a single, double E-type silhouette at 800 meters while wearing a protective mask.	Gunner is in stationary tripod firing position and engages a 800-meter, double E-type silhouette. Gunner will use 5-to 7-round bursts for this engagement.	14 rounds of 12.7-mm.	Gunner must impact one burst on the 800-meter target within 20 seconds.

Figure C-7. NBC Qualification Table III.

TASK	CONDITIONS TARGET/SITUATION	AMMO	STANDARD
3. Engage a single, double E-type silhouette at 400 meters while wearing a protective mask.	Gunner is in stationary tripod firing position and engages a 400-meter, double E-type silhouette. Gunner will use 5-to 7-round bursts for this engagement.	14 rounds of 12.7-mm.	Gunner must impact one burst on the 400-meter target within 20 seconds.
4. Engage a single, double E-type silhouette at 700 meters while wearing a protective mask.	Gunner is in stationary tripod firing position and engages a 700-meter, double E-type silhouette. Gunner will use 5-to 7-round bursts for this engagement.	14 rounds of 12.7-mm.	Gunner must impact one burst on the 700-meter target within 20 seconds.
5. Engage a single, double E-type silhouette at 1,000 meters while wearing a protective mask.	Gunner is in stationary tripod firing position and engages a 1,000-meter, vehicle type silhouette. Gunner will use 5-to 7-round bursts for this engagement.	14 rounds of 12.7-mm.	Gunner must impact one burst on the 1,000 meter target within 25 seconds.
6. Engage multiple, double E-type silhouettes at 400 and 700 meters while wearing a protective mask.	Gunner is in stationary tripod firing position and engages 400-and 700-meter, double E-type silhouettes. Gunner will use 5-to 7-round bursts for this engagement.	28 rounds of 12.7-mm.	Gunner must impact one burst on each target within 35 seconds.
7. Engage multiple, double E-type silhouettes at 550 and 800 meters while wearing the protective mask.	Gunner is in stationary tripod firing position and engages 550-and 800-meter, double E-type silhouettes. Gunner will use 5- to 7-round bursts for this engagement.	28 rounds of 12.7-mm.	Gunner must impact one burst on each target within 35 seconds.
8. Engage multiple, double E-type silhouettes at 400, 500, and 1,000 meters while wearing the protective mask.	Gunner is in stationary tripod firing position and engages scaled 400-,500-, and 1,000-meter, double E-type and vehicle type silhouettes. Gunner will use 5-to 7-round bursts for this engagement.	42 rounds of 12.7-mm.	Gunner must impact one burst on each target within 45 seconds.

Figure C-7. NBC Qualification Table III, continued.

C-5. NIGHT-FIRE EXERCISE

The night-fire exercise gives the soldiers the practical application of engaging targets using the AN/TVS-5 at night or during limited visibility.

a. **Objectives.** The objectives of this training are:

- To teach the gunners to zero the AN/TVS-5 to the M2.
- To engage targets at different ranges using the AN/TVS-5.
- To detect targets using the AN/TVS-5.

b. **Organization.** Organization of a unit for night firing is conducted the same as day fire. AN/TVS-5 zeroing is the first firing task of the night-fire phase. The night-fire portion of the course is also a requirement to become a qualified M2 gunner.

c. **Ammunition Issue.** The ammunition for the firing of this exercise will be broken down at the ammunition point and issued to each assistant gunner as he is assigned a firing point. The total ammunition for this exercise is 180 rounds, and issue will consist of the following rounds or belts:

- Task 1, 12 single rounds and two 7-round belts
- Task 2 through Task 8, one 154-round belt.

d. **Firing Sequence.** Firing of Table IV (Figure C-8) will be conducted in the following manner; these procedures pertain to a single gunner using a tripod-mounted M2 with an AN/TVS-5 If there is a second gunner, he will use the same procedures.

(1) *Task 1, Zero the AN/TVS-5.*

(a) The gunner must first mount the AN/TVS-5 to the weapon and place it into operation. Once the device is mounted, the gunner will fire a 7-round burst to seat the device, then fire another 7-round burst and tighten the device to make sure it is settled.

(b) The gunner must then center the reticle pattern in the field of view of the device. He then places the reticle aiming point on the 50-meter zero target aim point and fires three single rounds.

NOTE: Each click of the azimuth or elevation adjustment actuator moves the strike of the round 1/2 inch at 50 meters. One click of adjustment moves the reticle one square of the target at 50 meters.

(c) If the impact of the round (beaten zone) is over the target, the gunner has probably fired with an incorrect sight picture (too high on the target). He should re-fire concentrating on the proper point of aim and AN/TVS-5 sight picture. If the gunner again fails to zero with a proper device sight picture, he should inspect the sight calibration of the device.

(d) The gunner will observe the beaten zone and make adjustments to the sights so the strike of the round impacts on the target.

(e) After adjustments are made, the gunner will then fire another round at the same target until zero is obtained.

NOTE: If the gunner is able to zero his weapon using 12 rounds, he is then ready to continue the course. If not, he is removed from the firing line and given remedial training.

(2) *Task 2 through Task 8, Single and Multiple Engagements.*

(a) The gunner will load one 154-round belt of ammunition and fire bursts of 5 to 7 rounds at double E-type silhouettes at ranges of 800, 400, 700, and a 1,000-meter vehicular target. He will also engage multiple, double E-type silhouettes targets at 400 to 700; 550 and 800; and 400, 500, and 1,000 meters, respectively.

(b) The gunner will observe the beaten zone and make adjustments as necessary to hit the targets.

(c) After firing is completed, weapons will be cleared and the gunner critiqued. If there is a second gunner to fire, the entire procedure will be repeated.

NOTE: After both gunners have fired for practice, they will then be assigned new lanes and issued more ammunition for record firing. If possible, the gunner should keep the same weapon he used for practice fire. The only change in the above procedures will be in the last step; here, the gunner will be scored along with the critique. If the gunner is unable to meet standard, he is removed from the firing line, given remedial training, and re-fired at the commander's discretion.

DANGER

Before any gunner changes lanes, the lane NCO must ensure that the weapon has been properly cleared. Anyone observing any unsafe acts should immediately call a cease fire and notify range personnel.

e. **Scoring Procedures.** One point is given for each target hit with an additional 2-point bonus if the target is hit on the first burst. The total possible points for night fire is 33. A minimum 7 points (7 out of 11 exposures) is required to meet standard on the night-fire course.

NOTE: Firing Table IV is set up for gunners to engage targets out to 1,000 meters under moonlight. However, if visibility is limited by other conditions, then the commander may use his discretion to alter the ranges for better visibility.

TASK	CONDITIONS TARGET/SITUATION	AMMO	STANDARD
1. Zero the AN/TVS-5 to the M2 machine gun.	Gunner is in stationary tripod firing position at 50 meters and engages an M16A1/A2 zero target. Gunner will use single shots for this engagement. Two bursts will be used to spot the device.	12 single 12.7-mm rounds and two 7-round belts.	Gunner must impact 4 of 6 rounds of two 3-round shot groups within a 4-cm circle IAW FM 23-26.
2. Engage a single, double E-type silhouette at 800 meters while using the AN/TVS-5.	Gunner is in stationary tripod firing position and engages an 800-meter, double E-type silhouette. Gunner will use 5-to 7-round bursts for this engagement.	14 rounds of 12.7-mm.	Gunner must impact one burst on the 800-meter target within 20 seconds.
3. Engage a single, double E-type silhouette at 400 meters while using the AN/TVS-5.	Gunner is in stationary tripod firing position and engages a 400-meter, double E-type silhouette. Gunner will use 5-to 7-round bursts for this engagement.	14 rounds of 12.7-mm.	Gunner must impact one burst on the 400-meter target within 20 seconds.
4. Engage a single, double E-type silhouette at 700 meters while using the AN/TVS-5.	Gunner is in stationary tripod firing position and engages a 700-meter, double E-type silhouette. Gunner will use 5-to 7-round bursts for this engagement.	14 rounds of 12.7-mm.	Gunner must impact one burst on the 700-meter target within 25 seconds.
5. Engage a single, double E-type silhouette at 1,000 meters while using the AN/TVS-5.	Gunner is in stationary tripod firing position and engages a 1,000-meter, double E-type silhouette. Gunner will use 5- to 7-round bursts for this engagement.	14 rounds of 12.7-mm.	Gunner must impact one burst on the 1,000-meter target within 25 seconds.
6. Engage multiple, double E-type silhouettes at 400 and 700 meters while using the AN/TVS-5.	Gunner is in stationary tripod firing position and engages 400-and 700-meter, double E-type silhouettes. Gunner will use 5- to 7-round bursts for this engagement.	28 rounds of 12.7-mm.	Gunner must impact one burst on each target within 35 seconds.
7. Engage multiple, double E-type silhouettes at 550 and 800 meters while using the AN/TVS-5.	Gunner is in stationary tripod firing position and engages 550- and 800-meter, double E-type silhouettes. Gunner will use 5- to 7-round bursts for this engagement.	28 rounds of 12.7-mm.	Gunner must impact one burst on each target within 35 seconds.
8. Engage multiple, double E-type silhouettes at 400, 550, and1,000 meters while using the AN/TVS-5.	Gunner is in stationary tripod firing position and engages 400-, 550-, and 1,000-meter, double E-type silhouettes. Gunner will use 5- to 7-round bursts for this engagement.	42 rounds of 12.7-mm.	Gunner must impact one burst on each target within 45 seconds.

Figure C-8. Night Fire Qualification Table IV.

* C-6. QUALIFICATION STANDARDS

To qualify on the M2, the gunner must achieve the minimum standards on the firing tables.

a. **Firing Table I**. On this table, a gunner can earn one point for each round that hits in each space. He can earn up to 119 points on this table, but must earn at least 12 points (12 out of 17 pasters).

b. **Firing Tables II, III, and IV**. On each of these tables, the gunner can earn one point for each target hit plus a two-point bonus if he hits each target with the first burst. He can score up to 33 points on each of these tables, but to qualify, he must score at least 23 points on each.

c. **All Tables**. The gunner firing tables I through IV can earn up to 218 points. However, he must earn at least 12 points on Table I plus at least 23 points on each of the other three firing tables. Thus, he needs at least 12 + (3 x 23) = 12 + 69 = 81 points to qualify:

FIRING TABLE	POINTS NEEDED TO QUALIFY
Firing Table I	12
Firing Table II	23
Firing Table III	23
Firing Table IV	+ 23
	81 TOTAL POINTS NEEDED TO QUALIFY ON M2

d. **Ratings**. The grader rates each gunner as follows:

MAXIMUM.............................218
EXPERT...............................196 TO 217
FIRST CLASS.......................174 TO 195
SECOND CLASS..................153 TO 173
UNQUALIFIED......................152 OR LESS

e. **Scorecard**. This change prescribes a new edition of DA Form 7007-R. This blank, reproducible form may be copied from the back of this manual onto 8 1/2 by 11-inch paper. It is also available on the Army Electronic Library (AEL) CD-ROM (EM0001) and at the USAPA website located at (http://www.usapa.army.mil).

MACHINE GUN SCORECARD FOR M2

For use of this form, see FM 23-65, Change 1, or the back of this form. The proponent agency is TRADOC.

DATA REQUIRED BY THE PRIVACY ACT OF 1974

AUTHORITY: 10 USC 3012(g)/Executive order 9397.

PRINCIPAL PURPOSE: To aid individual training on targets at various ranges.

ROUTINE USE: To evaluate individual proficiency. SSN is used for positive identification purposes only.

DISCLOSURE: Voluntary. However, mass rating and scoring require some tracking method. Therefore, only those soldiers who provide their SSNs can receive scores or qualify on weapons or equipment.

1a. LAST NAME	1b. FIRST NAME	1c. MI	2. DATE *(YYYYMMDD)*
LEMAY	CURTIS	X	20021211

3. SSN	4. UNIT	5. LANE
222-22-2222	2/29TH INF	4

EXAMPLE

	6a. FIRING TABLE I			7a. FIRING TABLE II					8a. FIRING TABLE III					9a. FIRING TABLE IV				
TSK	RANGE (M)	HIT	PTS	RANGE (M)	TIME	HIT	PTS	BON	RANGE (M)	TIME	HIT	PTS	BON	RANGE (M)	TIME	HIT	PTS	BON
1	10	NA	NA	550	None	NA	NA	NA	550	None	NA	NA	NA	550	None	NA	NA	NA
2	10	NA	NA	800	20 Sec	X	1	2	800	20 Sec	X	1	2	800	20 Sec	X	1	2
3	10	NA	NA	400	20 Sec	X	1	2	400	20 Sec	X	1	0	400	20 Sec	X	1	2
4	10	NA	NA	700	25 Sec	X	1	2	700	25 Sec	X	1	2	700	25 Sec	X	1	2
5	10	X	20	1,000	25 Sec	X	1	2	1,000	25 Sec	X	1	2	1,000	25 Sec	X	1	2
6	10	X	30	400	35 Sec	X	1	2	400	35 Sec	X	1	2	400	35 Sec	X	1	2
				700		X	1	2	700		X	1	2	700		X	1	2
7	10	X	45	550	35 Sec	X	1	2	550	35 Sec	X	1	0	550	35 Sec	X	1	2
				800		X	1	2	800		X	1	2	800		X	1	2
8	NA	NA	NA	400	45 Sec	X	1	2	400	45 Sec	X	1	2	400	45 Sec	X	1	2
				550		X	1	0	550		X	1	2	550		X	1	2
				1,000		X	1	2	1,000		X	1	2	1,000		X	1	2

6b. SUBTOTAL FIRING TABLE I	7b. SUBTOTAL FIRING TABLE II	8b. SUBTOTAL FIRING TABLE III	9b. SUBTOTAL FIRING TABLE IV
95	31	29	33

10. TOTAL SCORE	11. RATING
188	GUNNER FIRST CLASS
12a. OIC's PRINTED OR TYPED NAME	**13a. GRADER'S PRINTED OR TYPED NAME**
SSG DWIGHT X. EISENHOWER	SSG EDWIN X. ROMMEL
12b. OIC'S SIGNATURE	**13b. GRADER'S SIGNATURE**
Dwight X. Eisenhower	Edwin X. Rommel

DA FORM 7007-R, JAN 2002 — DA FORM 7007-R, JUN 1991, IS OBSOLETE. — USAPA V1.00

Figure C-9. Example completed DA Form 7007-R (front).

The following procedure will be used to fill out the scorecard.

1. NAME: Enter last name , first name, and middle initial.
2. SSN: Enter gunner's social security number.
3. UNIT: Enter unit designation.
4. DATE: Enter date of firing.
5. LANE: Enter the lane number for the firing point of your gunner.
6. HIT: When firing Table I, enter the number of rounds impacting within the scoring space in task numbers 5, 6, and 7. For Tables II, III, and IV, enter an X when target is hit.
7. POINTS: Enter 1 point for each round impacting within the scoring space in task numbers 5, 6, and 7 in Table I. Gunners will receive 1 point for each target hit in Tables II, III, and IV.
8. BONUS: Entry is used only for Tables II, III, and IV. Two bonus points are awarded when gunner hits the target on initial burst of each target exposure.
9. TOTAL SCORE: Enter the total of combined hits, points, and bonus scores.

NOTE: It is possible that a gunner may score the required points outlined for qualification. However the gunner must also score 70 percent on each firing table before he is considered a qualified gunner.

MAXIMUM	218
EXPERT	196-218
GUNNER FIRST CLASS	174-195
GUNNER SECOND CLASS	153-173
UNQUALIFIED	152 and below

Back, DA Form 7007-R, Jun 91

Figure C-10. Example DA Form 7007-R, back.

Section II. MULTIPURPOSE RANGE COMPLEX

The multipurpose range complex is designed to conduct armor and mobilized machine gun firing. Detailed setup and target configuration is described in FM 25-7; its layout is shown in Figure C-11. This section also discusses the mounted firing exercise, requirements for crews, rating procedures, and standards.

C-7. DESIGN SPECIFICS

The tower and administrative facilities must be positioned so they do not impede tactical maneuver onto or off of the range.

a. Hardened emplacements are for aerial gunnery exercises in addition to tank and BFV gunnery.

b. System calibration targets must be provided at ranges of 950, 1,200, and 1,500 meters, as shown in Figure C-11.

c. Gunnery tasks that require the use of dud-producing ammunition cannot be fired on the range proper. Provisions for these tasks must be made in impact areas adjacent to the range.

d. Double target mechanisms are recommended for use in the first six target groupings for BFV training, or for positioning throughout the range as needed to support local training requirements.

e. The administrative area is not shown in exact location or scale.

f. Additional defilade positions maybe required for BFV gunnery.

g. The addition of battle positions and positioning of personnel targets closer to the baseline may be necessary to permit BFV dismounted infantry training.

Carrier 2 will upload and wait at the ammunition point. When the course is clear, carrier 2 will move to the initial fire point followed by carrier 4 and proceed as stated. Carrier 3 should now be at the ammunition point for uploading.

(1) *Task 1, Conduct Prefire Inspection.*

(a) The crew will move forward to the initial firing position. Once the carrier is in position, the crew will perform a prefire inspection by setting the headspace and timing, and load a 7-round belt of ammunition.

(b) The gunner will then fire the 7-round belt at a 450-meter target to ensure the machine gun is operational.

NOTE: If the machine gun fails to fire or fires sluggishly, recheck the headspace and timing and attempt to fire again. If the situation continues, clear the weapon and have the carrier move off the firing course and report to the armorer to have the problem corrected.

(c) Upon completion of Task 1, the gunner will clear the weapon and await further instructions.

NOTE: To accomplish Task 2 and Task 3, the gunner will load one 28-round belt of ammunition.

(2) *Task 2, Engage Stationary Target From a Mounted Stationary Firing Position.*

(a) The gunner will engage a 450-meter stationary personnel target from the initial firing location.

(b) The gunner will use a 5- to 7-round burst for this engagement.

(c) The gunner will observe the beaten zone and make adjustments as necessary to hit the target.

(d) Upon completion of Task 2, the gunner will await further instructions.

(3) *Task 3, Engage Stationary Target From a Stationary Firing Position.*

(a) From the initial firing location, the gunner will engage an 850-meter stationary vehicle target.

(b) The gunner will use a 5- to 7-round burst for this engagement.

(c) The gunner will observe the beaten zone and make adjustments as necessary to hit the target.

(d) Upon completion of Task 3, the gunner will await further instructions.

NOTE: On command from the grader, the crew will move the carrier to the next predetermined firing location, load one 28-round belt, and conduct Task 4.

(4) *Task 4, Engage Moving and Stationary Targets From a Mounted Stationary Firing Position.*

(a) From the second predetermined firing location, the gunner will engage one moving vehicle target at a range of 800 meters and one stationary vehicle target at a range of 1,000 meters.

(b) The gunner will use a 5-to 7-round burst for this engagement.

(c) The gunner will observe the beaten zone and make adjustments as necessary to hit the targets.

(d) Upon completion of firing Task 4, the gunner will await further instructions.

NOTE: When conducting Task 5, the gunner will load one 28-round belt and engage targets from a moving carrier along a predetermined route.

(5) *Task 5, Engage Stationary Targets While Firing From a Moving Carrier.*

(a) The carrier will be moving along the predetermined firing route, at which time the gunner will engage stationary vehicle targets at ranges of 300 and 500 meters.

(b) The gunner will use a 5- to 7-round burst for this engagement.

(c) The gunner will observe the beaten zone and make adjustments as necessary to hit the targets.

(d) Upon completion of Task 5, the gunner will await further instructions.

NOTE: On command from the grader, the crew will move the carrier to the third predetermined firing location, load one 28-round belt, and conduct Task 6.

(6) *Task 6, Engage a Stationary and a Moving Target From a Mounted Stationary Firing Position.*

(a) From the third predetermined firing location, the gunner will engage one 600-meter personnel target and one 800-meter moving vehicle target from a stationary firing position.

(b) The gunner will use a 5- to 7-round burst for this engagement.

(c) The gunner will observe the beaten zone and make adjustments as necessary to hit the targets.

(d) Upon completion of Task 6, the gunner will await further instructions.

NOTE: From the same firing location as Task 6, the crew will be placed in a chemical environment and be required to go to MOPP4.

(7) *Task 7, React in Chemical Environment.*

(a) Upon command, the crew will be required to react to a chemical agent.

(b) Upon completion of Task 7, the gunner will await further instructions.

NOTE: To accomplish Task 8 through Task 10, the gunner will load one 84-round belt of ammunition.

(8) *Task 8, Engage Stationary Target From a Mounted Stationary Firing Position While in MOPP4.*

(a) From the same firing location as Task 6, the gunner will engage a stationary target at a range of 500 meters from a mounted stationary firing position while in MOPP4.

(b) The gunner will use a 5- to 7-round burst for this engagement.

(c) The gunner will observe the beaten zone and make adjustments as necessary to hit the target.

(d) Upon completion of Task 8, the gunner will await further instructions.

(9) *Task 9, Engage Moving and Stationary Targets From a Mounted Stationary Firing Position While in MOPP4.*

(a) From the same firing location, the gunner will engage one 800-meter moving vehicle target and personnel targets at 1,000 meters.

(b) The gunner will use a 5- to 7-round burst for this engagement.

(c) The gunner will observe the beaten zone and make adjustments as necessary to hit the targets.

(d) Upon completion of Task 9, the gunner will await further instructions.

(10) *Task 10, Engage Multiple Stationary Targets in a Mounted Stationary Firing Position While in MOPP4.*

(a) From the same firing location, the gunner will engage one 300-meter personnel target and two stationary vehicle targets at ranges of 500 and 700 meters.

(b) The gunner will use a 5- to 7-round burst for this engagement.

(c) The gunner will observe the beaten zone and make necessary adjustments to hit the target.

(d) Upon completion of firing Task 10, the gunner will clear his machine gun, have it inspected by the safety officer, move from the firing line, and turn in any excess ammunition. Then the crew will be critiqued.

TASK	CONDITIONS TARGET/SITUATION	AMMO	STANDARD
1. Conduct prefire inspection.	At a stationary position given a 450-meter stationary target, headspace and timing gauge, unit's organic vehicle, and M2 machine gun.	7-round belt of 12.7-mm.	Conduct prefire inspection and fire 7 rounds to confirm operation of machine gun.
2. Engage stationary target from a stationary firing position.	Gunner is in a stationary mounted firing position and engages a 450-meter stationary personnel target. Gunner must use 5- to 7-round bursts for this engagement.	14 rounds of 12.7-mm.	Gunner must impact on target.
3. Engage stationary target from a stationary firing position.	Gunner is in a stationary mounted firing position and engages a 850-meter stationary vehicle. Gunner must use 5- to 7-round bursts for this engagement.	14 rounds of 12.7-mm.	Gunner must impact on target.
4. Engage moving and stationary targets from another stationary firing position.	Gunner is in a stationary mounted firing position and engages a moving vehicle at 800 meters and stationary vehicle at 1,000 meters. Gunner must use 5- to 7-round bursts for this engagement.	28-round belt of 12.7-mm.	Gunner must impact on each target.
5.Engage stationary targets while firing from a moving vehicle.	While the gunner's vehicle is moving, the gunner engages stationary vehicles at 300 and 500 meters. gunner must use 5- to 7-round bursts for this engagement.	28-round belt of 12.7-mm.	Gunner must impact on each target.
6. Engages a stationary and moving target from a stationary firing position.	Gunner is in a stationary firing position and engages a 600-meter personnel and 800-meter moving vehicle targets. gunner will use 5- to 7-round bursts for this engagement.	28-round belt of 12.7-mm.	Gunner must impact on each target.
7. React to chemical agent.	While in a stationary position, given protective mask and protective clothing, crew must react to chemical agent.	N/A	In accordance with FM 3-3.
8. Engage stationary target from a stationary firing position while in an upgraded NBC posture.	Gunner is in a stationary firing position and engages a 500-meter vehicle target. gunner will use 5- to 7-round bursts for this engagement.	14 rounds of 12.7-mm.	Gunner must impact on target.

Figure C-12. Mounted Firing Exercise Table VI.

TASK	CONDITIONS TARGET/SITUATION	AMMO	STANDARD
9. Engage moving and stationary targets while in a stationary firing position in an upgraded NBC posture.	Gunner will engage 800-meter moving vehicle target and personnel targets at 1,000 meters. Gunner will use 5- to 7-round bursts for this engagement.	28 rounds of 12.7-mm.	Gunner must impact on targets.
10. Engage multiple stationary targets while in a stationary firing position in an upgraded NBC posture.	Gunner will engage 300-meter personnel targets, 500-and 700-meter vehicle targets. Gunner will use 5- to 7-round bursts for this engagement.	42 rounds of 12.7-mm.	Gunner must impact on each target.

Figure C-12. Mounted Firing Exercise Table VI, continued.

e. **Scoring Procedures.** The mounted firing exercise is graded on a Go/No Go basis. Commanders may increase the difficulty of any or all tasks to align the exercise with the unit's mission.

C-9. PREDETERMINED FIRING EXERCISE

The predetermined firing exercise is for the gun crew that has demonstrated proficiency during the basic phase of gunnery. Emphasis will be on developing range cards and confirming range-card data during day and night firing.

a. **Objectives.** The objective of this training is to reinforce what was developed in the fundamental gunnery phases. It is designed to increase the effectiveness of the M2 MG crew by building their confidence to quickly and accurately deliver a large volume of fire on a prescribed target.

b. **Organization.** The unit is assembled in the bleachers, given instructions, and briefed on training that will be conducted while they are on the range. After briefing, they will be organized into gun crews and moved to firing lanes. Lanes will be used IAW local range policies.

c. **Ammunition.** This exercise requires 168 rounds of 12.7-mm linked ammunition. The gunner is allotted two bursts per target during the day phase and one burst per target during the night phase. Each gunner will be issued two belts of ammunition; one 112-round belt for the day phase and one 56-round belt for the night phase.

d **Firing Sequence.** The sequence of firing will be conducted IAW Firing Table VII (Figure C-13, page 33). The suggested sequence of firing is as follows:

(1) *Day phase.*

(a) Task 1, Prepare a Range Card. Once the gunner is assigned his firing point, he must prepare a range card for that position. Range cards must be prepared IAW paragraph E-2.

(b) Task 2, Obtain Direction and Elevation Readings for the Final Protective Line. During this task, the gunner is allotted 28 rounds to obtain and record the direction and elevation reading of his FPL.

(c) Task 3, Obtain Direction and Elevation Readings for Point Targets. The gunner will be required to obtain and record direction and elevation readings for point targets located at 400, 600, 800, and 1,000 meters. He is allotted 56 rounds for this task.

(d) Task 4, Obtain Direction and Elevation Readings for Linear Targets. The gunner will be required to obtain and record direction and elevation readings for linear targets located at 600 and 800 meters. He is allotted 28 rounds for this task.

(2) *Night phase.*

NOTE: Each crew is scored as a group in the night phase. (The day phase is not scored.) The crew applies the data obtained during the day and engages their targets. Each crew receives 10 points for each target engagement. A minimum of 40 out of a possible 70 is required. This exercise is not a requirement for qualification; however, commanders can use this training to test their gun crews' proficiency.

(a) Task 5, Engage Point Targets Using Range-Card Data. The gunner must engage point targets located at 400, 600, 800, and 1,000 meters using range-card data. He is allotted 28 rounds for this task. The grader will announce the sequence of engagements.

(b) Task 6, Engage Area Targets Using Range-Card Data. The gunner must engage linear targets located at 600 and 800 meters using range-card data. He is allotted 14 rounds for this task. The grader will announce the sequence of engagements.

(c) Task 7, Fire Final Protective Line. The gunner will fire his FPL to obtain grazing fire. He is allotted 14 rounds for this task.

e. **Conduct of Firing.** The gunner, assistant, and leader will prepare a range card for that position. Once complete, each crew will be given 168 rounds as prescribed in a predetermined firing table.

(1) Each gunner initially lays on his target by using the dry-fire technique. Each crew is issued 168 rounds of ammunition, and the gunner is ordered to load by the group NCO. When all the gunners of a group are ready, the NCO announces, "Up," to the officer in charge.

(2) When all groups are ready to fire, the OIC announces, "Give me an 'Up' when you have engaged all targets and have obtained the correct data to all targets. You are clear to fire."

(3) When the gunner has correctly engaged his target (FPL), the leader records the information from the T&E mechanism and traversing bar onto a range card. Then, the assistant gunner becomes the gunner, the gunner becomes the leader, and the leader becomes the assistant gunner.

(4) As each member of the crew becomes the gunner, he fires at a different preselected target (order number 2, a linear target; order number 3, a point target).

(5) After each crew member has been a gunner, assistant gunner, and leader and all data on the three targets have been obtained, the group NCO clears and checks the machine guns of his group and announces to the OIC, "Group cleared and checked."

TASK	CONDITIONS TARGET/SITUATION	AMMO	STANDARD
1. Prepare a range card for the M2 machine gun.	The gunner is given a range card, a selected firing position, M2 machine gun, M3 tripod, and silhouette targets located from 400 to 1,000 meters.	N/A	Gunner must prepare a range card IAW Appendix E, FM 23-65.
2. Determine and record direction and elevation for final protective line.	The gunner is allotted 12.7-mm linked ammo, and a sector of fire with limits.	28 rounds of 12.7-mm.	Gunner must record the direction and elevation for his range card.
3. Determine and record direction and elevation for point targets.	The gunner is allotted 12.7-mm linked ammo, four E-type silhouettes located at 400, 600, 800, and 1,000 meters.	56 rounds of 12.7-mm.	Gunner must impact one burst on each target.
4. Determine and record direction and elevation for area targets.	The gunner is allotted 12.7-mm linked ammo, two E-type silhouettes located at 600 and 800 meters.	28 rounds of 12.7-mm.	Gunner must impact one burst on each target.
5. Engage point targets using range card data.	The gunner is allotted 12.7-mm linked ammo, the same firing position, and his prepared range card.	28 rounds of 12.7-mm.	Gunner must impact one burst on each target.

Figure C-13. Predetermined Firing Table VII.

TASK	CONDITIONS TARGET/SITUATION	AMMO	STANDARD
6. Engage area targets using range card data.	The gunner is allotted 12.7-mm linked ammo, and his prepared range card.	14 rounds of 12.7-mm.	Gunner must impact one burst on each target.
7. Fire final protective line.	The gunner is allotted 12.7-mm linked ammo, and his prepared range card.	14 rounds of 12.7-mm.	Gunner must obtain grazing fire when firing his FPL.

Figure C-13. Predetermined Firing Table VII, continued.

APPENDIX D

FIGHTING POSITIONS

Fighting positions are stable platforms from which a gunner can engage the enemy. Fighting positions are dug as often as possible, but for the most part only when a unit is in the defense and has the time. This appendix discusses the construction of only two fighting positions: the one-, two-, or three-man, tripod-mounted position; and the M63 antiaircraft emplacement.

D-1. CONSTRUCTION CONSIDERATIONS

When constructing a fighting position, the leader and gunner should consider factors that affect the position, such as cover, concealment, fields of fire, size, and improvements.

a. The cover in a machine gun fighting position protects the gun crew from frontal small-arms fire and from fragments of high explosive shells impacting within 3 to 5 meters of the position. The frontal parapet must be at least one M16 rifle length thick and high enough to hide the helmets of the soldiers in the position. Overhead cover built out of sandbags stacked on logs will protect from shell fragments. The logs should be at least 6 inches thick, with at least two layers of sandbags over them (Figure D-1).

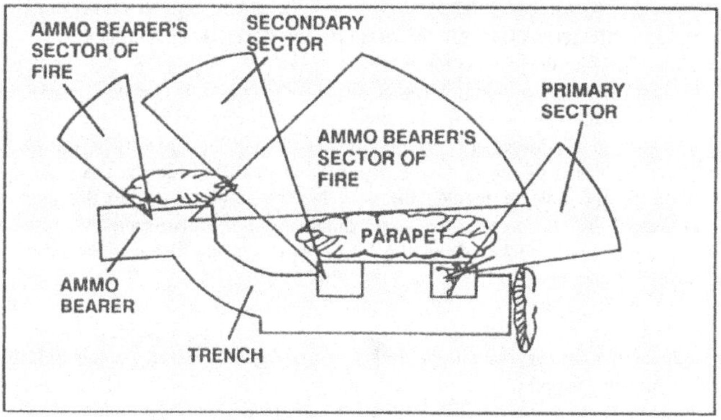

Figure D-1. Three-man fighting position.

b. Concealment hides the machine gun fighting position from direct observation by the enemy, who will make every attempt to locate and destroy heavy machine gun positions early in the battle. The position must be made to blend with its surroundings with both natural and man-made camouflage. Excess dirt from the initial digging should be moved away from the position, and the sod used to re-cover the overhead protection. From the enemy side, the position should not be visible at 35 meters or less (hand grenade range). It should never be obvious even at close range. Use sod or dampen the ground in front of the muzzle to reduce the dust cloud caused by firing. Consider what the position looks like from the air and camouflage it to blend with the surrounding terrain.

c. Selective clearing that does not destroy natural camouflage may be necessary to provide good observation and fields of fire in both primary and secondary sectors of fire.

d. The position should be armpit deep, be wide enough to allow two soldiers with load-bearing equipment to move freely, and have two distinct firing platforms.

e. Improvements to the position may include a grenade sump, a sloping floor with shallow trench for drainage, and a rear parapet for protection against shell fragments and small-arms fire from the rear (other friendly positions or supporting fire from armored personnel carriers).

D-2. CONSTRUCTION OF A TRIPOD-MOUNTED POSITION

After being assigned a sector of fire with an FPL or a PDF, the gun crew begins constructing the fighting position. The tripod is placed in position first and marked so that the weapon will be pointed in the general direction of the target area. A preliminary sketched range card is drawn to show the limits of the sector. The gun crew then outlines the shape of the platform and hole to include the area for the frontal cover in the ground (Figure D-2). The crew then starts digging out the platforms. When they get about 4 to 6 inches deep, the MG is put in place to cover the primary sector of fire until construction is complete.

a. When assigned an FPL, emplace the gun by locking the traversing slide to the extreme left or right of the traversing bar, depending on which side of the primary sector the FPL is on. Align the barrel on the FPL by shifting the tripod.

NOTE: No direction entry is needed in the data section of the range card for the FPL.

b. When assigned a PDF, emplace the gun by locking the traversing slide at the center of the traversing bar. Shift the tripod and gun until the barrel is aimed at the center of the sector. Check coverage of the sector limits by traversing the gun fully left and right.

Figure D-2. Planning the fighting position.

NOTE: In the data section of the range card, record the direction and elevation data of the PDF and the sector limits from the T&E mechanism.

c. The crew digging the hole uses the dug-up dirt to build up the cover – first for frontal cover and then for flank and rear cover. They dig the hole deep enough to protect the crew and still let the gunner shoot with comfort (usually about armpit deep) (Figure D-3). They fix the tripod legs in place by digging, sandbagging, or staking them down. This will ensure that the gun does not shift during firing, which would render the range card data useless.

Figure D-3. Digging the fighting position.

(1) The crew digs three trench-shaped grenade sumps at various points where the crew can kick grenades into them (Figure D-4).

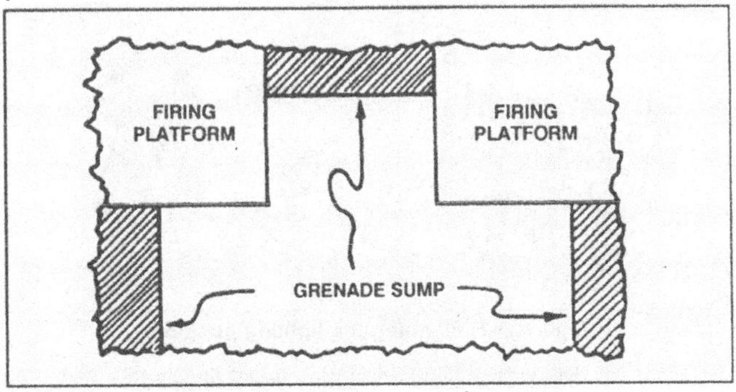

Figure D-4. Digging grenade sumps.

(2) When a position does not have a secondary sector of fire, the crew digs only half of the position (Figure D-5).

Figure D-5. Half of a position.

(3) When a position has both a primary and secondary sector, the crew prepares two firing platforms. The crew prepares overhead cover for a machine gun position like that of a two-man small arms fighting position. Time and material permitting, overhead cover should extend to cover the firing platforms (Figure D-6). Failure to properly construct overhead cover can result in reduced fields of fire, inability to mount NVDs, or problems in reloading. Proper construction of overhead cover is critical to survival.

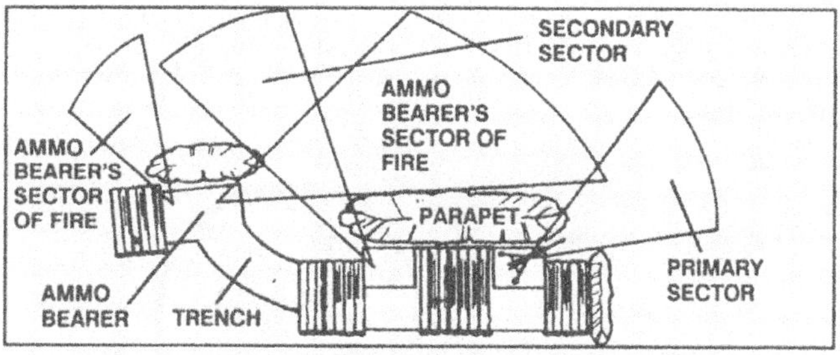

Figure D-6. Two firing platforms with overhead cover.

(4) When there is a three-man crew for the machine gun, the ammunition bearer digs a one-man fighting position to the flank of the gun position so that he can see and shoot to the oblique. This will allow him to cover the front of the machine gun's position (Figure D-7).

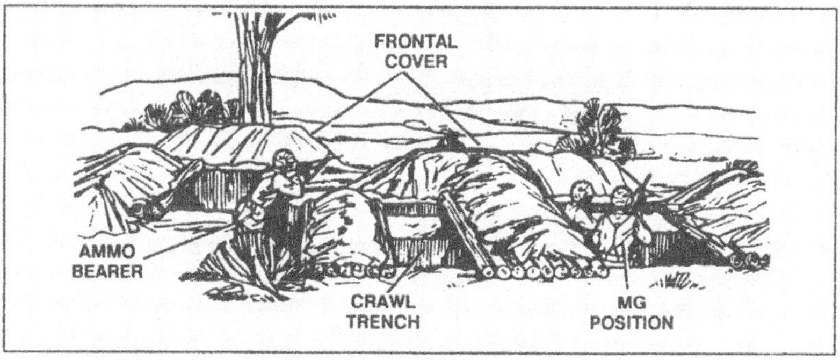

Figure D-7. Ammo bearer covering the front.

D-3. CONSTRUCTION OF AN ANTIAIRCRAFT EMPLACEMENT

Place the machine gun on the M63 mount alongside the designated location of the gun position. If needed, it can fire at any time during the construction process. Construct a circular position, with sufficient room to traverse the gun 360 degrees (normally, three and a half M16 lengths in diameter and one M16 length deep). Use the dirt dug out of the circular hole to construct a wall of sandbags around the position. Using sandbags, build a stable platform for the mount. Stake down the legs of the mount and place sandbags over them. This will prevent the mount from tipping backward when low-level targets are engaged. Include the M63-mounted machine gun's fire in the plan for defending the unit against ground attack (Figure D-8).

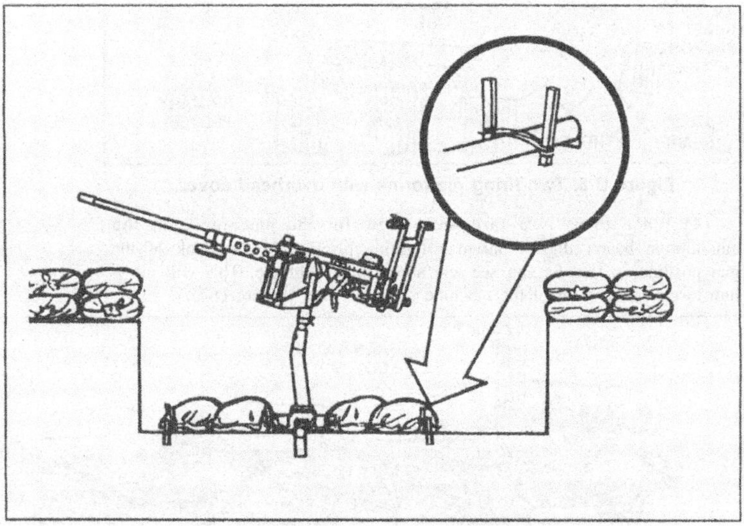

Figure D-8. Open pit-type antiaircraft position.

APPENDIX E

RANGE CARDS

A range card is a sketch or diagram of the terrain that a weapon is assigned to cover by fire. It shows possible target areas and terrain features plotted in relation to a firing position. The information on a range card is used for planning and controlling fire, for rapidly detecting and engaging targets, and for orienting replacement personnel or units. Therefore DA Form 5517-R, should be used to record the information.

E-1. RANGE CARD SECTIONS

Each machine gunner normally prepares range cards for his fighting position. He prepares one for each primary, alternate, and supplementary position designated in the defense and for any static position when enemy contact is possible; for example, a position in an assembly area. Each range card contains, as a minimum, the following information:

a. The symbol for the weapon covering the sector.

b. The azimuth (degrees) and distances (meters) of the firing position from an easily recognizable terrain feature. (This serves as an easy reference to locate the firing position.) If there is no easily recognizable terrain feature, an eight-digit grid maybe used.

c. The boundaries of the area assigned to be covered by observation and fire.

d. Areas where targets are likely to appear (engagement areas) and the range, azimuth, and elevation to them from the firing position.

e. Dead space (areas that cannot be observed or covered by fire).

f. The direction of magnetic north when the range card is properly oriented.

g. Identification data.

- Unit designation (no higher than company).
- Time and date of preparation.
- Firing position (primary, alternate, or supplementary).

E-2. PREPARATION

The following steps are taken to prepare a range card. (See Figure E-1, page E-3, for an example of a completed range card).

a. Draw the symbol for the MG in the lower center of the range card.

b. Show the sector of fire by drawing solid lines from the weapon symbol to the left and right limits. Sketch any easily recognizable terrain features that can be used to identify the sector. In the data section, indicate the magnetic azimuth and the range to the far limits of the sectors of fire.

NOTE: The left and right limits are labeled one and two, respectively, in the sector sketch and a circle drawn around each number. The left and right limits should be the first items drawn and labeled in the sector sketch and the data section of the range card. (The azimuth and range are not placed on the solid lines for left or right limit.)

c. Place target reference points at the location designated by the platoon leader or unit SOP, and at any other locations where a target is likely to appear. Number each TRP and likely target locations in the sketch section of the range card.

d. Draw a maximum engagement line across the sector of fire for the MG. These lines depict the maximum ranges at which a target can be effectively engaged.

e. Show dead space or areas where targets cannot be engaged with direct fire by drawing diagonal lines across the areas and writing the words DEAD SPACE.

f. Draw a magnetic north arrow on the range card to orient it with the terrain. Then add identification data – unit designation (no higher than company level), time and date of preparation, and type of position (primary, alternate, or supplementary).

g. Enter the information for the weapon reference point in the remarks block on the range card.

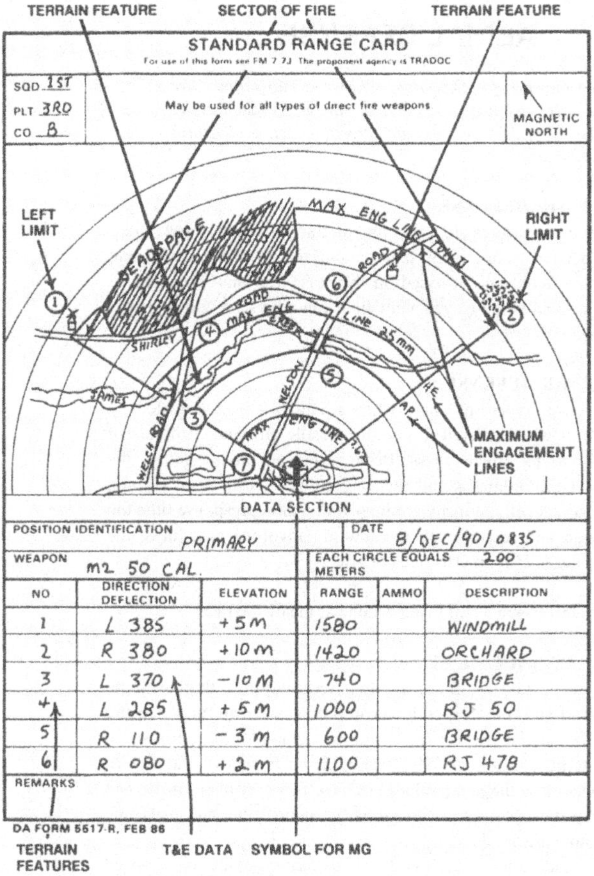

NO	DIRECTION DEFLECTION	ELEVATION	RANGE	AMMO	DESCRIPTION
1	L 385	+5M	1580		WINDMILL
2	R 380	+10m	1420		ORCHARD
3	L 370	-10M	740		BRIDGE
4	L 285	+5m	1000		RJ 50
5	R 110	-3m	600		BRIDGE
6	R 080	+2m	1100		RJ 478

Figure E-1. Range card.

* APPENDIX F

AERIAL DEFENSE

The two methods for defending against aerial attacks are passive air defense and active air defense. Passive measures are taken to avoid air detection; active measures combat air attacks.

F-1. PASSIVE AIR DEFENSE

Passive air defense measures are a first line of defense. They include troop or vehicle dispersion, concealment and camouflage, and observation and reporting. Attack avoidance is based on the reasoning that what can be seen from the air can be destroyed and what cannot be seen will probably not be attacked.

F-2. ACTIVE AIR DEFENSE

Active air defense comes into play when a unit has been detected by hostile aircraft or when ordered to interdict hostile aircraft. Volume of fire is the key to its effectiveness. Because the action is so fast, the response must be coordinated and tightly controlled, usually at platoon level. At company level, too many are involved, making response time too slow. At section and squad level, too few are involved to achieve the volume of fire needed for an effective defense.

NOTE: All aircraft engagement firing is done at the cyclic rate.

F-3. AERIAL ENGAGEMENT

In engaging aerial targets, there are two requirements that must be considered: lead and line. Superelevation (compensation made for the pull of gravity on the projectile) is another consideration for some weapons, but the caliber .50 MG projectile is basically flat out to 800 meters. Therefore the gunner does not have to worry about it. He can instead observe the tracers crossing the target course line and make the required adjustment from them.

a. **Lead.** Leading the aircraft is the compensation made for its speed when aiming the weapon so the projectile and the aircraft will arrive at the same spot at the same time. The required lead for a given aircraft is always calculated as a certain number of lengths of that aircraft. A high performance aircraft will need a greater lead factor than a slow one. The gunner should be taught to estimate leads by projecting images of the fuselage in the front of the aircraft.

(1) If the gunner is able to track an aircraft and lead the air craft with a constant number of leads, there will be two points along the target course line where he will attain hits. As long as the gunner understands this principle, he can bring effective fire on an aircraft.

* (2) There are many different types of hostile aircraft (FM 44-80). To simplify things for the gunner, all aircraft are classified into two types: high performance and low performance. Any aircraft that has an estimated speed of more than 150 knots is classified as high performance and anything slower is low performance.

b. **Line.** Line is the requirement that the projectile must intersect the target course line of the aircraft. If the projectiles do not pass through this target course line, the aircraft cannot be engaged.

F-4. FLY-THROUGH TECHNIQUE

As an aerial target moves along its target course line, the lead required to engage the target changes because the range and angle between the aircraft and the gun position are changing. The lead required increases from the initial sighting (A in Figure F-1) to midpoint (B in Figure F-1). The required lead is at its maximum at midpoint.

a. As the aircraft moves beyond the midpoint on the target course line, the lead needed begins to decrease. Therefore, by leading an aerial target by less than the maximum required lead, there will be two points along the target course line where the constant lead will be the correct lead and the aircraft will fly through the projectiles.

b. In instructing the fly-through method of engagement, do not teach the gunners to bring continuous fire from points A through E, teach them to bring fire at points A through B and again before point D through the second fly through. As the gunner becomes proficient in this method, he will learn to adjust his lead and bring longer and more accurate fire on the target.

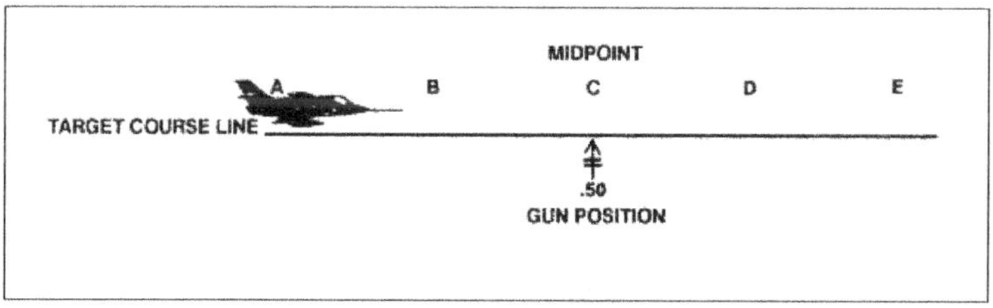

Figure F-1. Aerial target engagement.

* APPENDIX G

NIGHT OPTICS

Weapons-mounted night optics are less versatile than goggles with regard to situational awareness. However, they offer greater magnification and resolution at crew-served weapons ranges.

The graduated reticles on night optics offer the flexibility of engaging at various ranges. Laser pointers fix on one range, so the firer must "aim off" before engaging targets at a different range.

Night optics weigh more than goggles--they are awkward to move with and employ. Using night optics requires that the firer make specific adjustments to firing positions. Also, night optics might work poorly in some ambient light or thermal conditions.

Despite their drawbacks, properly boresighted night optics used in the right conditions offer units an extraordinary option: to engage targets beyond the range of opposing force weapons at night.

Section I. AN/PAS-13 (V3) HEAVY WEAPON THERMAL SIGHT

The AN/PAS-13 (V3) heavy weapon thermal sight (HWTS) (Figure G-1, page G-2) is a silent, lightweight, compact, durable, battery-powered, infrared imaging sensor that consumes little battery power. The self-contained infrared (IR) imaging sensor in this sight helps the firer acquire the target in low visibility conditions.

G-1. EQUIPMENT DESCRIPTION AND OPERATION

The HWTS works well at night and in the daytime. The telescope's IR sensor receives infrared light, converts it to digital data, processes it, and then displays it digitally as an infrared image for the user.

a. **Components**. The HWTS has two functional groups: the basic sensor and the telescope. See Figure G-1, page G-2, for equipment data.

(1) ***Basic Sensor***. The scanner reflects the IR light it receives from the telescope onto the detective (sensor) assembly. The assembly senses IR light, converts it into a video image, then conditions the video for display on the LED array. The LED array illuminates both the IR image and the reticle. This light image reflects off the scanner, which forms the actual image the firer sees in the eyepiece.

(2) ***Telescope***. The telescope receives IR light from an intended target and its surroundings, and then magnifies and projects this light onto the scanner on the basic sensor.

b. **Compatibility**. The heavy weapon thermal sight fits the weapons shown in Figure G-2:

- M2 .50-caliber machine gun.
- M24 sniper rifle.
- MK 19 grenade machine gun.
- Squad leader's M16.

c. **Operational Modes**. To place the HWTS in operational mode--

(1) If you have an AN/PAS-13 (V3), see TM 11-5855-302-12&P.

(2) If you have an AN/PAS-13 (OMNI), see TM 11-5855-312-10.

Weight:	4.5 pounds
Field of View (FOV):	
Narrow	3 degrees
Wide	9 degrees
Telescopic Magnification:	
Narrow FOV	10X
Wide FOV	3.3X
Dimensions:	
Length:	18 inches
Width:	6.25 inches
Height:	6.25 inches
Power Consumption:	
ON mode	12 watts
STANDBY mode	5.5 watts
Battery Life:	
71% STANDBY mode	10 hours
EMERGENCY mode	3.5 hours

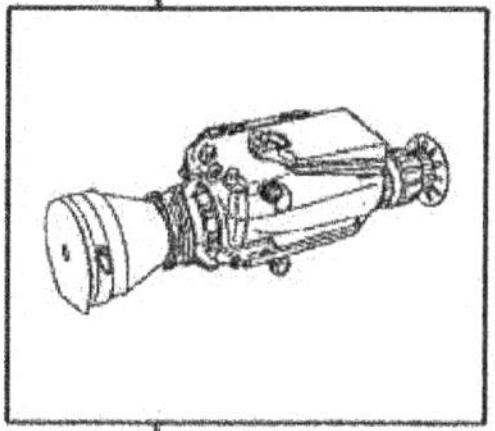

Figure G-1. Heavy weapon thermal sight.

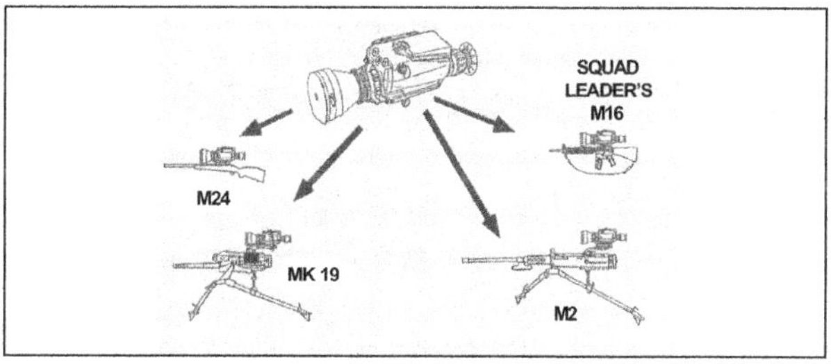

Figure G-2. Weapons that use the AN/PAS-13 (V3) HWTS.

G-2. FUNDAMENTALS OF MARKSMANSHIP

The fundamentals of marksmanship remain about the same for the AN/PAS-13 (V3) heavy weapon thermal sight (HWTS) as they do for the

AN/PVS-4. (TM 11-5855-302-12&P discusses HWTS reticle selection and point of aim.)

a. **Firing Position.**

(1) ***Sitting***. When using the tripod in the low or high position, sit directly behind the gun between the trail legs of the tripod. Extend your legs under the tripod, cross them, or brace your feet on the tripod. Regardless of which you choose, place your elbows on the inside of your thighs for support. Place your right eye in the HWTS eyepiece.

(2) ***Standing***. With the M2 mounted on a vehicle, stand with both hands on the control grips and your thumbs resting on the trigger. Keep your elbows against your body, your body forward, and your chest against your hands to brace the gun. Place your right eye in the HWTS eyepiece.

(3) ***Kneeling***. When using the M2 in a fighting or hasty tripod-mounted position, kneel and grasp the control grips with your thumbs on the trigger. Place your eye in the HWTS eyepiece.

(4) ***Prone***. Use the prone position when firing from a tripod that is sitting in a low position. Lie on the ground directly behind the gun and spread your legs a comfortable distance apart, with your toes pointed outward. Rest your left elbow on the ground and grasp the elevating handwheel of the T&E mechanism with your left hand. Grasp the right spade grip with your right hand, with the right thumb in position to press the trigger.

b. **Aiming**.

(1) Determine the range to the target. The aiming point of the M2 reticle depends on the range to the target. To determine range, you can use a range card, estimate range, use the reticle's aiming box, or use TRPs. The HWTS has two M2 reticles--a wide and a narrow field of view.

(2) Use the wide or narrow field of view to scan, to help determine range, and to engage targets.

(3) Use the indicators on the reticle to determine range, or use other methods of range determination.

(4) Set the mounting bracket range selector to NEAR for any target less than 1,400 meters and to FAR for any target beyond 1,400 meters.

(5) Using the T&E mechanism, sight the reticle aiming point on the target (bottom center).

(6) Take the slack out of the T&E by holding the M2 down and to one side. A loose T&E mechanism can move the HWTS reticle off the target's aiming point.

NOTE: Squad leader: observe the impact of the round to help the gunner adjust the sight to bring the strike of the round onto the target.

c. **Breath Control**. This fundamental of marksmanship does not change.

d. **Trigger Squeeze**. This fundamental of marksmanship does not change.

> **WARNING**
>
> ***Do not* press your eye hard against the HWTS eyepiece while engaging a target. The M2 kicks back slightly and could harm you. Secure the M2 as well as you can to increase accuracy and to help prevent injury.**

e. **Boresighting**. This paragraph provides guidance on boresighting. For zeroing procedures, see TM 11-5855-302-12&P and TM 11-5855-312-10. (This stays in the pocket of the sight's soft storage case.) Boresight as follows:

(1) Place the M2 in the ready-to-fire position 10 meters from the boresight target offset. To avoid a failed boresight zero, level the target and weapon before you adjust the laser.

NOTE: The HWTS will not focus in on a silhouette copied to plain paper. To create a boresight target offset, use a 25-meter M16 zero target and attach to it a copy of the boresight target offset.

(2) Mount the HWTS and boresight the weapon. Set the M2 mount to the NEAR position.

(3) Use the reticle select switch to change to the M2 reticle. The reticle type displays in the upper right corner on the raster.

(4) Make adjustments with the T&E until the bore light moves onto the bore light aiming point on the boresight target offset.

(5) Adjust the reticle with the reticle adjuster on the right side of the HWTS. Remember to hold the adjuster down for a count of five. Press and hold the adjuster up to move the reticle up, down to move the reticle down, forward to move the reticle left, and back to move the reticle right.

(6) Use the vertical and horizontal reticle adjuster to move the 1,000-meter aiming point of the reticle to the HWTS aiming point on the boresight target offset.

(7) Recheck the bore laser point of aim and make sure the reticle remains in the center mass of the HWTS aim point on the boresight target offset.

TRAINING STRATEGIES:	When using night optics, M2 gunners use night initial training strategies. Additional nonfiring exercises also apply.

f. **Qualifications**. Changes to M2 qualifications accommodate new night vision sights and laser pointers. Using the new tables reduces the time and ammunition required to train and qualify gun crews. Lack of ranges and other resources previously limited qualification options. The new tables increase the opportunities for qualification and the flexibility in the process of qualification. Simpler scoring and grading procedures allow gunners to maintain a high level of proficiency.

(1) With the availability of night vision sights and laser pointers, the current night standards are not challenging. Taking into consideration the "own the night" concept, gunners need to qualify at night to the same or a greater standard than we do during the day. Using these proposed qualification tables and training strategies, the gunner can do so. Each night practice and qualification table has three variations, to reflect the different range, weapon, and sight combinations. Each day table has a variation for each type target.

(2) Use the training strategies and preliminary marksmanship instruction previously described to prepare for day practice and qualification tables. Graders need to prepare, and to ensure ranges are prepared, in accordance to the latest published manuals on the subject. After the gunner trains, he should use Firing Table I then Firing Table II. If he fails to meet the standards in Firing Table I, he must repeat the training and fire the table again. Gunners *must* pass the practice table before moving on to the qualification table.

(3) The gunner must pass day qualification before proceeding to night qualification training.

g. **Nonfiring Exercise**. Use this exercise to train the gunner to use the T&E in order to sight the laser on target.

(1) Place the M2 in the ready-to-fire configuration, with the sight mounted on the M2, and with the M2 mounted on the M3 tripod.

(2) Have the gunner close his left eye and look into the sight with his right eye.

(3) Tell the gunner to focus on a target in either the narrow or wide field of view.

(4) Tell the gunner to close both eyes. After he closes his eyes, you, the trainer, must turn and change the setting on the sight.

(5) Tell the gunner to achieve the same sight picture he had before you reset it.

(6) Look through the sight to ensure the gunner has acquired a clear sight picture. If not, show him how to get one.

(7) Repeat the exercise until the gunner achieves proficiency with the sight.

h. **T&E Manipulation and Aiming Exercise**. Use this exercise to train the gunner to manipulate the T&E, acquire targets, and properly sight the target based on the target range. This exercise also trains the gunner to use the mounting bracket to adjust for range.

(1) Set up two to three targets that the gunner can acquire with the HWTS using the T&E.

(2) Regardless of the actual range to the target, tell the gunner to sight in on the first target and give him the range you want it to be.

(3) Ensure the gunner sets the mounting bracket to the correct setting and uses the T&E correctly.

(4) Time the gunner as he conducts this exercise, and create a competition between the gunners in training.

i. **Practice Qualification.** Use practice qualification, modeled after the day qualification scenario, to transition the gunner into engaging multiple timed targets at various ranges.

j. **Qualification.** Use qualification to assess the gunner's ability to engage targets on a timed scenario. If he does not meet the standard during the night practice table, train him again. He must meet the standard on the practice table before firing the qualification table. Conduct the nonfiring exercise as follows:

(1) Place the M2, with HWTS mounted, in the ready-to-fire configuration on the M3 tripod.

(2) Have the gunner close his left eye and look into the HWTS with his right eye.

(3) Tell the gunner to focus on a target in either the narrow or wide field of view.

(4) Tell the gunner to close both eyes. After he closes them, change the setting on the sight.

(5) Tell the gunner to open his eyes and to achieve the same sight picture he had before you reset it.

(6) Look through the sight to ensure the gunner has acquired a clear sight picture. If not, show him how to get one.

(7) Repeat the exercise until the gunner reaches proficiency with the sight.

Section II. AN/TVS-5

Gunners use the AN/TVS-5 (Figure G-3, page G-7) a portable, battery-operated electro-optical instrument, to observe and aim weapons fire at night. The AN/TVS-5 amplifies reflected light such as moonlight, starlight, and sky glow, making the scene clearly visible to the operator. The sight *does not* emit visible or infrared light (except from the eyepiece) that the enemy can detect. By using this device, the gunner can observe the area, and then detect and engage any suitable target. The quality of the zero determines the usefulness of the weapon. Accurate zeroing requires practice. Figure G-3 also provides equipment data for the AN/TVS-5.

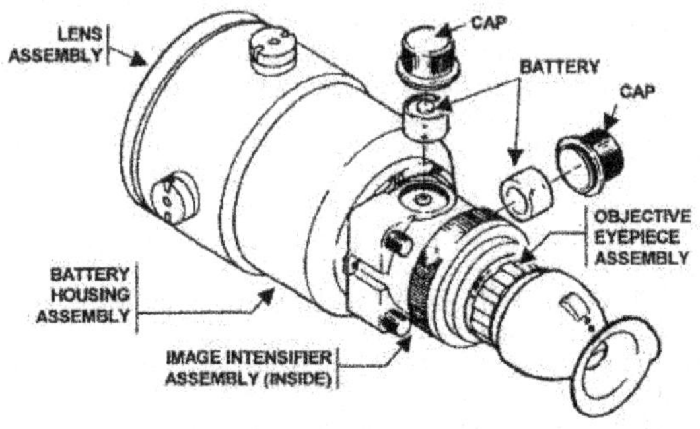

EQUIPMENT DATA

Magnification	5.6X
Field of view	9 degrees
Eyepiece focus	+3 to -6 diopters
Objective lens focus	25 meters to infinity
Range	1,000 to 1,200 meters for vehicle targets (under ideal conditions)
Reticle adjustment	+/-2.5 degrees (in ¼-mil increments)
Operating conditions	-65 F (-54C) to +125 F (+52C)
Battery type	BA-5567/U (2 each)
Voltage	3.0 volts DC
Shelf life	One year at 70 F (21C)
Operating life	About 50 hours at normal temperature

Figure G-3. AN/TVS-5 components and data.

CAUTION

Handle the AN/TVS-5, a precision electro-optical instrument, carefully at all times.

G-3. OPERATION

The AN/TVS-5 has the following controls and indicators:

a. Use the ON-OFF/TUBE BRIGHTNESS control to apply power to the sight and to control the brightness of the image-intensifier tube. This control also enables the ON-OFF RETICLE BRIGHTNESS control to function.

b. Use the ON-OFF RETICLE BRIGHTNESS to apply power to the reticle and to control the brightness of the reticle.

c. Use the OBJECTIVE FOCUS RING to adjust the range focus from 25 meters to infinity.

d. Use the DIOPTER FOCUS RING to focus the eyepiece.

e. Use the DIOPTER INDICATOR to learn the current direction of rotation of the DIOPTER FOCUS RING for plus and minus diopters.

f. Use the RETICLE ELEVATION ADJUSTMENT ACTUATOR to adjust the reticle up and down. Each click moves the strike of the round 1 inch at 100 meters.

g. Use the RETICLE AZIMUTH ADJUSTMENT ACTUATOR to adjust the reticle right or left. Each click moves the strike of the round 1 inch at 100 meters.

WARNING

- ***Do not*** **press the eye guard except with your eye area and then only to operate the sight. Used improperly, the sight emits illumination that the enemy can detect.**
- ***Do not*** **use the sight without the eye guard attached, or you may receive an injury when the weapon recoils.**
- **Use care when discarding the batteries. Their contents cause extreme irritation to the eyes and to oral and nasal passages. To prevent explosion, avoid burning the batteries.**
- ***Do not*** **short-circuit the batteries.**
- ***Do not*** **recharge the batteries.**
- **Remove batteries before storing the night vision sight.**
- **Always replace both mercury batteries at the same time.**

G-4. MOUNTING AND DISMOUNTING PROCEDURES

The AN/TVS-5 mounts and dismounts as follows:

a. **Mounting Procedures**. If the mounting bracket (Figure G-4) has not been installed already, install it now as described in TM 11-5855-214-10.

(1) Align the scribe line on the sight with the scribe line on the bracket.

(2) Place the sight in the groove of the bracket and tighten the lever screw clockwise.

(3) Secure the lever screw with lacing wire or tape to ensure the sight does not vibrate loose.

(4) Seat the device. After firing the initial burst, retighten the lever screw to ensure a secure mount for the sight. If unable to fire at this time, lightly shake the sight to ensure it is mounted correctly.

b. **Dismounting Procedures**.

(1) Remove the lacing wire or tape from the lever screw.

(2) Loosen the lever screw until the sight comes free, and then lift it off the bracket.

(3) Remove the batteries and place the sight in its carrying case.

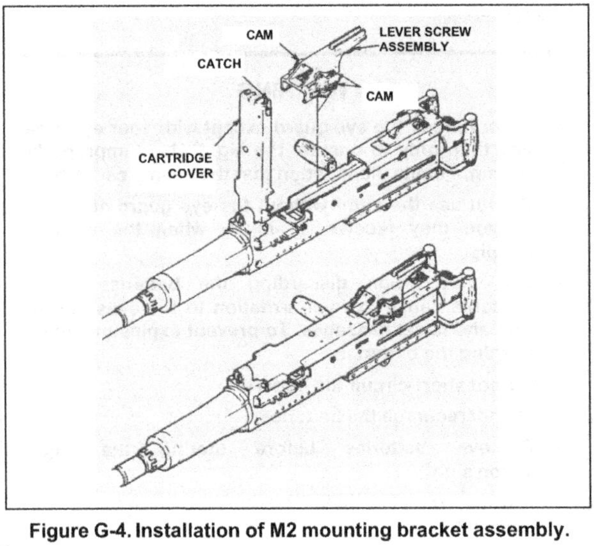

Figure G-4. Installation of M2 mounting bracket assembly.

G-5. FUNDAMENTALS OF MARKSMANSHIP

Except for aiming, the fundamentals of marksmanship for the AN/TVS-5 are the same as those for the AN/PAS-13 heavy weapon thermal sight.

a. Determine the range to the target (Figure G-5, page G-10). The range setting on the HWTS depends on the range to the target. To determine range to target, use any of several methods: range cards, range estimation techniques, the upper portion of the AN/TVS-5 reticle, or TRPs.

b. After determining range, use the aiming point designated for that range on the M2 reticle (Figure G-6, page G-12 shows the old reticle; Figure G-7, page G-13 shows the new reticle). If using the HWTS bracket and the M2 reticle, adjust the bracket to the nearest settings. Zero the sight on this setting and use the appropriate dot for the distance to the target as indicated in the reticle. For ranges beyond 1,200 meters, change the bracket to the far setting.

- Change "4" to 1,400 meters.
- Change "6" to 1,500 meters.
- Change "8" to 1,600 meters.
- Change "10" to 1,700 meters.
- Change "12" to 1,800 meters.

c. Bring the aiming point of the AN/TVS-5 onto the target using the T&E as explained in Section V.

d. Hold the weapon tight against the T&E, press the trigger, and adjust fire as needed.

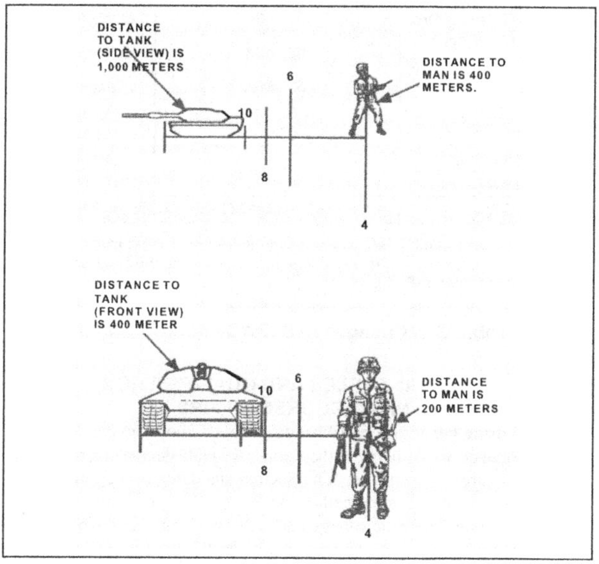

Figure G-5. Range estimation for M2 (old reticle).

G-6. BORESIGHTING PROCEDURES

Boresighting the AN/TVS-5 to the M2 requires the following steps:

a. Place the M2 in the ready-to-fire configuration, with the AN/TVS-5 mounted, 25 meters from the bore-light-offset zero targets. Ensure the weapon and target sit level before making any adjustments.

b. Adjust the T&E until the bore light moves onto the bore light aiming point on the boresight target offset.

c. Use the vertical and horizontal reticle adjusters to move the 1,000-meter aiming point on the reticle to the AN/TVS-5 aiming point.

d. Recheck the bore light aiming point and the AN/TVS-5 aiming points to ensure they remain center mass of the target.

G-7. AMMUNITION, NIGHT TRAINING STRATEGY

Table G-1 shows STRAC changes effective with this change publication.

EVENT	CURRENT STRAC BALL/MIX	REC'D CHGS	TOTAL BALL/MIX
10M ZERO	12 BALL	0	12 BALL
10M RECORD	119 BALL	0	119 BALL
TRANSITION ZERO, PRACTICE	50 MIX	ELIMINATE	0
TRANSITION RECORD	154 MIX	ELIMINATE	0
NIGHT ZERO, PRACTICE, RECORD	182 MIX	ELIMINATE	0
AG INSTR FIRE	54 MIX	ELIMINATE	0
TABLE I	0	140 MIX	140 MIX
TABLE II	0	140 MIX	140 MIX
TABLE III	0	140 MIX	140 MIX
TABLE IV	0	140 MIX	140 MIX
TOTAL	131 / 440 = 571	560 MIX	131/560 = 691

NOTE: The above STRAC supports one qualification. Allocations increase with the frequency of qualifications required. Check the firing unit's force activity designator (FAD).

Table G-1. Changes to STRAC effective with this publication.

Section III. TABLES AND QUALIFICATION, NIGHT COURSE OF FIRE

Improved firing tables and qualification methods allow the gunner to track his performance while he fires the tables. Requalifying gunners need not repeat the entire course of fire. This section standardizes the four tables to lead the gunner on a gradual path to qualification. The gunner must pass each firing table before he moves on to the next one. The tables in this section require 691 rounds to qualify a gunner--93 more than the old tables required, and 120 more than the STRAC previously authorized. The guidance in this appendix supersedes the STRAC. Though the new tables allow more rounds, they cut the time allowed to qualify. However, they recommend a band of ranges rather than a specific range for each target. This adds considerable flexibility. The ten-meter firing exercise is only a means of training--it is not a qualification table.

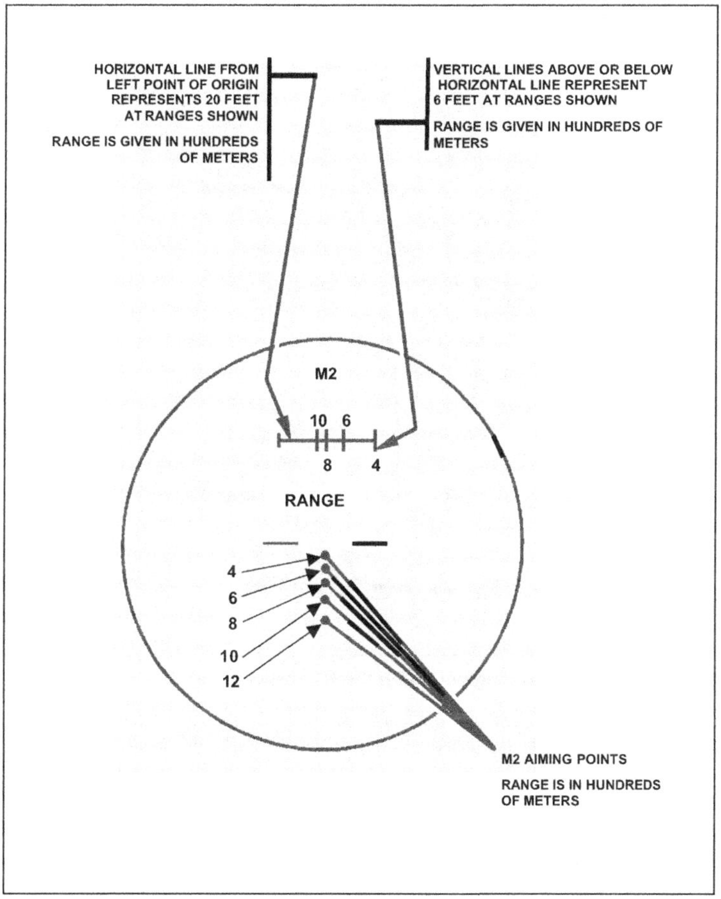

Figure G-6. Aiming points for the M2 (old reticle).

G-8. DIFFERENCES BETWEEN PRACTICE AND QUALIFICATION TABLES

Practice tables allow thirty extra seconds for each engagement. Also, infantry gun crews can practice in the daytime, but must qualify in both the daytime and at night.

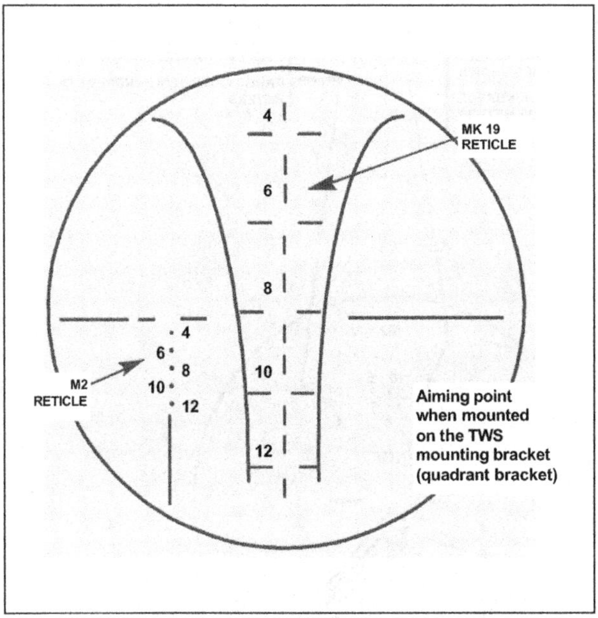

Figure G-7. Aiming points for the M2 (new reticle).

a. **Scoring**. For point target engagements (lightly armored vehicle targets, such as BRDMs and BTRs), give full credit (GO) when the gunner hits the target.

b. **Range Setup**. Set up targets and ranges as follows. Select target ranges IAW the scorecard for the applicable firing table. When choosing or placing targets, make sure no dead space falls within 100 meters of any of them. Any dead space near the target could keep the gunner from seeing the round impact, which he has to do in order to adjust fire:

(1) ***Hull Targets***. You need not modify hull targets on an impact range.

(2) ***Popup Silhouettes***. Set up a thermal source on popup silhouette targets so the gunner can acquire them with the HWTS. If the gunner will be using the AN/TVS-5, place a light source on each target.

- Targets between 400 and 600 meters--two chemical lights.
- Targets between 600 and 900 meters--three chemical lights.

c. **Grading**. Each firing point requires one grader.

(1) ***Grading Equipment***. While firing the day tables, the grader needs a set of binoculars. His other equipment needs vary according to the type of range used at night.

(a) *Night Vision Assistance*. When firing at hull targets, or when popup targets provide no downrange feedback, the grader must obtain an appropriate night vision device with which to observe the strike of the round.

- AN/PVS-14, 7B, with 3X magnifier.
- AN/TAS-4.
- AN/PAS-13 (heavy).

(b) *Target Scenario*. The grader needs this.

(c) *Range and Range Card*. At each particular firing point, the grader gives the gunner a target and a range at which the gunner must engage the target. The grader needs a range card for that firing range. The range card should include numbered targets and the ranges to them.

(2) ***Start and End Times***. Time starts as soon as the grader provides the target range and the target appears--which occur in this order, since graders must provide all required information before the target appears. Sometimes the targets consist of vehicle hulls in stationary positions. In this case, time starts as soon as the grader tells the gunner which target to engage and gives him the range to the target. Regardless of the type of target, time ends when the time designated for that task expires, or when the gunner successfully engages the target.

(3) ***Ammunition***. Ammunition breaks down by task. The gunner places each belt beside the assistant gunner in order of use. The number of rounds authorized for each task equals the number of rounds in each belt. For example, eight tasks require eight belts of ammunition, which should all sit within reach of the assistant gunner in the order the gunner plans to fire them.

(4) ***Fire Control***. Controlling M2 firing presents no problem when the range setup includes a firing lane for each firing point. However, this ideal situation seldom occurs with hull targets. In fact, some ranges must use the same target for each lane. This can cause confusion. When, inevitably, more than one gunner fires at the same target, no one can tell who hit it. The 400-meter target usually presents this problem. To prevent it, make sure that only one gunner at a time engages each target. If you want to let more than one gunner to fire at the same time, and then mix up the tasks so the gunners are firing at different targets. For example, have Point One fire at the 1,500-meter target while Point Two engages the 600-meter target, and so on. This requires careful coordination and communication between the graders.

d. **Firing Tables**. Figures G-8 and G-9 (page G-16), Figure G-10 (page G-17), and Figure G-11 (page G-18) show completed examples of the recommended day and night practice and qualification tables for the infantry M2 gunner and crew. Blank, reproducible forms are included at the back of the manual for users to copy onto 8 1/2 by 11-inch paper. They are also available on the Army Electronic Library (AEL) CD-ROM (EM0001) and at the USAPA website: http://www.usapa.army.mil.

(1) The first task allows evaluation of field zeroing. The gunner confirms the zero, even if he boresighted the weapon. If the gunner fails to zero within 14 rounds, graders remove him from the line and train him some more before letting him refire the table.

(2) All tasks have point targets. Graders should change one or two targets between the ranges of 600 and 900 meters to area targets, but should not change the round count or the time. They base their changes on range resources and the commander's guidance.

(3) Graders match the correct table to each target and weapon-sight configuration.

(4) Graders use range finders to predetermine the range to each target from each firing point.

(5) Graders set up M2s mounted or dismounted (in the tripod configuration), based on the range constraints and the commander's guidance.

(6) Leaders can have the unit fire the tasks in any order, and should do so to avoid redundancy in practice and qualification.

(7) For NBC qualification, the leader uses the practice tables.

e. **Ammunition Requirements for Night Training Strategy**. Table G-1 shows the ammunition requirements for a night training strategy.

G-9. SPECIAL MOUNTING PROCEDURES

This paragraph explains the special mounting and dismounting procedures required to mount a night vision device on the M2 heavy barrel .50 caliber machine gun.

a. **Mounting Procedure**. Before mounting the sight, the gunner ensures the bolt is forward and the rear sight is in the down position.

(1) Release and raise the top cover assembly to the upright position.

(2) Place the mounting bracket over the breach and slide it rearward until it stops.

(3) Push the three locking cams rearward until the bracket is secured (lock the side-locking cam first), and close the top cover assembly.

M2 CALIBER .50 HEAVY BARREL MACHINE GUN FIRING TABLE I
DAY PRACTICE SCORECARD

For use of this form, see FM 23-65, Change 1; the proponent agency is TRADOC.

DATA REQUIRED BY THE PRIVACY ACT OF 1974

AUTHORITY:	10 USC 3012(g)/Executive order 9397.
PRINCIPAL PURPOSE:	To aid individual training on targets at various ranges.
ROUTINE USES:	To evaluate individual proficiency. SSN is used for positive identification purposes only.
DISCLOSURE:	Voluntary. However, mass rating and scoring require some tracking method. Therefore, only those soldiers who provide their SSNs can receive scores or qualify on weapons or equipment.

1a. LAST NAME	1b. FIRST NAME	1c. MI	2. DATE *(YYYYMMDD)*
DOE	JOHN	A	20021201

3. SSN	4. UNIT	5. RANGE	6. LANE
222-22-2222	2 / 29 TH INF	RUTH	4

7a. ENGAGEMENT	7b. STANDARDS	7c. AMMO	7d. TIME *(Minutes)*	7e. RANGE *(Meters)*	7f. GO	7g. NO GO
ZERO	1 BURST HIT	14	NA	400	NA	NA
2	1 BURST HIT	14	1.5	1,100 (± 200)	X	
3	1 BURST HIT	14	1.5	1,500 (± 200)	X	
4	1 BURST HIT	14	1.5	600 (± 100)	X	
5	1 BURST HIT	14	1.5	800 (± 100)	X	
6	1 BURST HIT	14	1.5	400	X	
8. MULTIPLE TARGETS						
7	1 BURST HIT	28	2.5	1,100 (± 200)	X	
	1 BURST HIT			600 (± 100)	X	
8	1 BURST HIT	28	2.5	800 (± 100)	X	
	1 BURST HIT			1,500 (± 200)	X	

EXAMPLE

9. SCORING *(Check one)*

EXPERT - 9 [X] SHARPSHOOTER - 8 [ ]

MARKSMAN - 6 [ ] UNQUALIFIED - 5 OR BELOW [ ]

10. GUNNER'S SIGNATURE

PFC John A. Doe

11a. GRADER'S PRINTED OR TYPED NAME	11b. GRADER'S SIGNATURE
SSG DAVID R. JONES	*SSG David R. Jones*

DA FORM 7448-R, JAN 2002 USAPA V1.00

Figure G-8. Example completed DA Form 7448-R (Day Practice Scorecard).

M2 CALIBER .50 HEAVY BARREL MACHINE GUN FIRING TABLE II
DAY QUALIFICATION SCORECARD

For use of this form, see FM 23-65, Change 1; the proponent agency is TRADOC.

DATA REQUIRED BY THE PRIVACY ACT OF 1974

AUTHORITY: 10 USC 3012(g)/Executive order 9397.
PRINCIPAL PURPOSE: To aid individual training on targets at various ranges.
ROUTINE USES: To evaluate individual proficiency. SSN is used for positive identification purposes only.
DISCLOSURE: Voluntary. However, mass rating and scoring require some tracking method. Therefore, only those soldiers who provide their SSNs can receive scores or qualify on weapons or equipment.

1a. LAST NAME	1b. FIRST NAME	1c. MI	2. DATE *(YYYYMMDD)*
DOE	JOHN	A	20021201

3. SSN	4. UNIT	5. RANGE	6. LANE
222-22-2222	2 / 29 TH INF	RUTH	4

7a. ENGAGEMENT	7b. STANDARDS	7c. AMMO	7d. TIME *(Minutes)*	7e. RANGE *(Meters)*	7f. GO	7g. NO GO
ZERO	1 BURST HIT	14	NA	400	NA	NA
2	1 BURST HIT	14	1	1,100 (± 200)		X
3	1 BURST HIT	14	1	1,500 (± 200)	X	
4	1 BURST HIT	14	1	600 (± 100)	X	
5	1 BURST HIT	14	1	800 (± 100)	X	
6	1 BURST HIT	14	1	400	X	
8. MULTIPLE TARGETS						
7	1 BURST HIT	28	2	1,100 (± 200)	X	
	1 BURST HIT			600 (± 100)	X	
8	1 BURST HIT	28	2	800 (± 100)	X	
	1 BURST HIT			1,500 (± 200)	X	

EXAMPLE

9. SCORING *(Check one)*

EXPERT - 9 [ ]	SHARPSHOOTER - 8 [X]
MARKSMAN - 6 [ ]	UNQUALIFIED - 5 OR BELOW [ ]

10. GUNNER'S SIGNATURE

PFC John A. Doe

11a. GRADER'S PRINTED OR TYPED NAME	11b. GRADER'S SIGNATURE
SSG DAVID R. JONES	SSG David R. Jones

DA FORM 7449-R, JAN 2002 USAPA V1.00ES

Figure G-9. Example completed DA Form 7449-R (Day Qualification Scorecard).

M2 CALIBER .50 HEAVY BARREL MACHINE GUN FIRING TABLES III (A) AND III (B)
NIGHT PRACTICE SCORECARDS

For use of this form, see FM 23-65, Change 2; the proponent agency is TRADOC.

DATA REQUIRED BY THE PRIVACY ACT OF 1974

AUTHORITY: 10 USC 3012(g)/Executive order 9397.
PRINCIPAL PURPOSE: To aid individual training on targets at various ranges.
ROUTINE USES: To evaluate individual proficiency. SSN is used for positive identification purposes only.
DISCLOSURE: Voluntary. However, mass rating and scoring require some tracking method. Therefore, only those soldiers who provide their SSNs can receive scores or qualify on weapons or equipment.

1a. LAST NAME	1b. FIRST NAME	1c. MI	1d. RANK	2. SSN
Doe	John	A	PFC	222-22-2222

TABLE III (A). NIGHT PRACTICE SCORECARD

NOTE: Use this table if the M2 has the TWS, or if it has an AN/TVS-5 that has been upgraded with the third-generation tube.

a. RANGE	b. UNIT	c. LANE	d. DATE *(YYYYMMDD)*
Ruth	2 / 29 th	4	20021120

e. ENGAGEMENT	f. STANDARDS	g. AMMO	h. TIME *(Minutes)*	i. RANGE *(Meters)*	j. GO	k. NO GO
ZERO	1 BURST HIT	14	NA	400	NA	NA
2	1 BURST HIT	14	1.5	1,100 (±200)	X	
3	1 BURST HIT	14	1.5	1,500 (±200)	X	
4	1 BURST HIT	14	1.5	600 (±100)		X
5	1 BURST HIT	14	1.5	800 (±100)	X	
6	1 BURST HIT	14	1.5	400	X	
l. MULTIPLE TARGETS						
7	1 BURST HIT	28	2.5	1,100 (±200)	X	
	1 BURST HIT			600 (±100)	X	
8	1 BURST HIT	28	2.5	800 (±100)	X	
	1 BURST HIT			1,500 (±200)	X	

4. SCORING *(Check one)*
EXPERT - 9 [ ]
SHARPSHOOTER - 8 [X]
MARKSMAN - 6 [ ]
UNQUALIFIED - 5 OR BELOW [ ]

5. GUNNER'S SIGNATURE: PFC John A. Doe
6a. GRADER'S PRINTED OR TYPED NAME: SSG David R. Jones
6b. GRADER'S SIGNATURE: SSG David R. Jones

7. TABLE III (B). NIGHT PRACTICE SCORECARD

NOTE: Use this table if the M2 has an AN/TVS-5 that has not been upgraded.

a. RANGE	b. UNIT	c. LANE	d. DATE *(YYYYMMDD)*
Ruth	2 / 29 th	4	20021120

e. ENGAGEMENT	f. STANDARDS	g. AMMO	h. TIME *(Minutes)*	i. RANGE *(Meters)*	j. GO	k. NO GO
ZERO	1 BURST HIT	14	NA	400	NA	NA
2	1 BURST HIT	14	1.5	1,100 (±200)	X	
3	1 BURST HIT	14	1.5	800 (±100)	X	
4	1 BURST HIT	14	1.5	600 (±100)	X	
5	1 BURST HIT	14	1.5	800 (±100)	X	
6	1 BURST HIT	14	1.5	400	X	
l. MULTIPLE TARGETS						
7	1 BURST HIT	28	2.5	1,100 (±200)	X	
	1 BURST HIT			600 (±100)	X	
8	1 BURST HIT	28	2.5	800 (±100)	X	
	1 BURST HIT			600 (±100)	X	

8. SCORING *(Check one)*
EXPERT - 9 [X]
SHARPSHOOTER - 8 [ ]
MARKSMAN - 6 [ ]
UNQUALIFIED - 5 OR BELOW [ ]

9. GUNNER'S SIGNATURE: PFC John A. Doe
10a. GRADER'S PRINTED OR TYPED NAME: SSG David R. Jones
10b. GRADER'S SIGNATURE: SSG David R. Jones

DA FORM 7450-R, OCT 2002 DA FORM 7450-R, JAN 2002, IS OBSOLETE. USAPA V1.00ES

EXAMPLE

Figure G-10. Example completed DA Form 7450-R (Night Practice Scorecard).

M2 CALIBER .50 HEAVY BARREL MACHINE GUN FIRING TABLES IV (A) AND IV (B)
NIGHT QUALIFICATION SCORECARDS

For use of this form, see FM 23-65, Change 2; the proponent agency is TRADOC.

DATA REQUIRED BY THE PRIVACY ACT OF 1974

AUTHORITY:	10 USC 3012(g)/Executive order 9397.
PRINCIPAL PURPOSE:	To aid individual training on targets at various ranges.
ROUTINE USES:	To evaluate individual proficiency. SSN is used for positive identification purposes only.
DISCLOSURE:	Voluntary. However, mass rating and scoring require some tracking method. Therefore, only those soldiers who provide their SSNs can receive scores or qualify on weapons or equipment.

1a. LAST NAME	1b. FIRST NAME	1c. MI	1d. RANK	2. SSN
Doe	John	A	PFC	222-22-2222

3. TABLE IV (A). NIGHT QUALIFICATION SCORECARD

NOTE: Use this table if the M2 has the TWS, or if it has an AN/TVS-5 that has been upgraded with the third-generation tube.

a. RANGE	b. UNIT	c. LANE	d. DATE *(YYYYMMDD)*
Ruth	2 / 29	4	20021204

e. ENGAGEMENT	f. STANDARDS	g. AMMO	h. TIME *(Minutes)*	i. RANGE *(Meters)*	j. GO	k. NO GO
ZERO	1 BURST HIT	14	NA	400	NA	NA
2	1 BURST HIT	14	1	1,100 (± 200)	X	
3	1 BURST HIT	14	1	1,500 (± 200)	X	
4	1 BURST HIT	14	1	600 (± 100)		X
5	1 BURST HIT	14	1	800 (± 100)	X	
6	1 BURST HIT	14	1	400	X	
l. MULTIPLE TARGETS						
7	1 BURST HIT	28	2	1,100 (± 200)	X	
	1 BURST HIT			600 (± 100)		X
8	1 BURST HIT	28	2	800 (± 100)	X	
	1 BURST HIT			1,500 (± 200)	X	

4. SCORING *(Check one)*	
EXPERT - 9	
SHARPSHOOTER - 8	
MARKSMAN - 6	X
UNQUALIFIED - 5 OR BELOW	

5. GUNNER'S SIGNATURE: PFC John A. Doe

6a. GRADER'S PRINTED OR TYPED NAME: SSG David R. Jones

6b. GRADER'S SIGNATURE: SSG David R. Jones

7. TABLE IV (B). NIGHT QUALIFICATION SCORECARD

NOTE: Use this table if the M2 has an AN/TVS-5 that has not been upgraded.

a. RANGE	b. UNIT	c. LANE	d. DATE *(YYYYMMDD)*
Ruth	2 / 29	4	20021204

e. ENGAGEMENT	f. STANDARDS	g. AMMO	h. TIME *(Minutes)*	i. RANGE *(Meters)*	j. GO	k. NO GO
ZERO	1 BURST HIT	14	NA	400	NA	NA
2	1 BURST HIT	14	1	1,100 (± 200)	X	
3	1 BURST HIT	14	1	800 (± 100)	X	
4	1 BURST HIT	14	1	600 (± 100)	X	
5	1 BURST HIT	14	1	800 (± 100)	X	
6	1 BURST HIT	14	1	400	X	
l. MULTIPLE TARGETS						
7	1 BURST HIT	28	2	1,100 (± 200)		X
	1 BURST HIT			600 (± 100)	X	
8	1 BURST HIT	28	2	800 (± 100)	X	
	1 BURST HIT			600 (± 100)	X	

8. SCORING *(Check one)*	
EXPERT - 9	
SHARPSHOOTER - 8	X
MARKSMAN - 6	
UNQUALIFIED - 5 OR BELOW	

9. GUNNER'S SIGNATURE: PFC John A. Doe

10a. GRADER'S PRINTED OR TYPED NAME: SSG David R. Jones

10b. GRADER'S SIGNATURE: SSG David R. Jones

DA FORM 7451-R, OCT 2002 DA FORM 7451-R, JAN 2002, IS OBSOLETE. USAPA V1.00ES

EXAMPLE

Figure G-11. Example completed DA Form 7451-R (Night Qualification Scorecard).

DANGER

BEFORE INSTALLING THE BRACKET, MAKE SURE THE WEAPON IS *UNLOADED* AND THE SAFETY IS SET ON *SAFE*. A LOADED WEAPON CAN DISCHARGE ACCIDENTALLY, INJURING OR KILLING SOMEONE.

NOTE: If the bracket sticks before reaching its correct position, rock the bracket up and down while sliding it.

b. **Dismounting Procedure**. Ensure the weapon is clear and the bolt is forward.

(1) Raise the top cover assembly to the upright position.

(2) Push the three locking cams forward until the bracket is free.

(3) Slide the mounting bracket forward over the breach until the mounting bracket clears the weapon.

Section IV. BORE-LIGHTING PROCESS AND BORESIGHT TARGET OFFSETS

The bore light allows gunners to accurately zero weapons and most aided vision equipment without the use of bullets. Table G-2 shows different combinations of weapon and aided-vision device that gunners can use to zero the bore light.

DANGER

- ***DO NOT STARE*** **INTO THE VISIBLE LASER BEAM.**
- ***DO NOT LOOK*** **INTO THE VISIBLE LASER BEAM THROUGH A TELESCOPE OR A PAIR OF BINOCULARS.**
- ***DO NOT POINT*** **THE VISIBLE LASER BEAM AT A MIRROR-LIKE SURFACE.**
- ***DO NOT SHINE*** **THE VISIBLE LASER BEAM INTO OTHER PEOPLE'S EYES.**

WARNING

- **Make sure the weapon is CLEAR and on SAFE before using the bore light.**
- **When rotating the bore light to zero it, ensure the mandrel turns. *Do not* tighten the bore light down on the mandrel; doing so could strip or break the mandrel's stud.**

	M16A2	M4/MWS	M249	M240B/M60	M2	M203	AT4
AN/PAQ-4C	X	X	X	X		X	X
AN/PEQ-2A	X	X	X	X	X	X	X
AN/PAS-13	X	X	X	X	X		X
M68 CCO	X	X					
AN/PVS-4	X	X	X	X		X	
AN/TVS-5					X		

Table G-2. Possible weapon-aided vision-device combinations.

G-10. BORESIGHTING PROCESS

Properly boresighting a weapon requires two gunners--a firer and a target-holder.

a. **Personnel.**

(1) *Firer*. The firer zeroes the bore light and adjusts the aided vision device. Before boresighting, he locks the M2 down in a cradle or some other device to keep the weapon stable during the boresighting process. In the absence of such a device, he assumes the most stable supported firing position possible and makes sure the weapon does not cant during boresighting.

(2) *Target holder*. The target holder keeps the target straight up and down at the appropriate distance from the firer and directs the firer in making necessary sight adjustments. The target holder *must* wear night vision goggles when viewing laser aiming light offsets.

b. **Steps**.

(1) *Align Bore Light*. Checking the alignment of the bore light requires the following actions:

(a) Place the appropriate mandrel, with bore light attached, in the muzzle of the weapon.

(b) Turn the bore light on so that the laser beam strikes the target (offset) 10 meters away.

(c) Slowly rotate the bore light one-half turn (180 degrees) while watching the beam on the target area, noting any circular patterns made.

(d) If the beam remains stationary, you have boresighted the M2. Move forward to the next paragraph and use the appropriate boresight target for the given weapon.

(e) If the beam rotates in a circle, adjust the windage, elevation, or both, until the beam remains stationary or rotates on itself no more than 1 centimeter.

(2) *Adjust Bore Light*. If necessary, the firer and target holder must adjust the bore light.

(a) Move the target 2 meters away.

(b) Mark the location of the laser beam.

(c) Slowly rotate the bore light one-half turn.

(d) Note the new location of the laser beam.

(e) Adjust the windage and elevation until the laser beam moves halfway back to its original location.

(3) *Adjust elevation adjustment screw*. One click at 25 meters equals 4 millimeters; clockwise equals down.

(4) *Adjust windage adjustment screw*. One click at 25 meters equals 4 millimeters; clockwise equals left.

(5) *Continue adjusting the bore light*. Keep adjusting it until the laser beam either stops moving or spins upon itself within 1 centimeter.

(6) *Move the target, and recheck the boresight*. Move it 10 meters away and recheck the boresight. Repeat this process every 10 meters if necessary.

NOTE: Because the M2 requires such a large offset, place the offset for the boresight target 5 meters away.

(7) *Boresight the Aided Vision Device to the Weapon.* Select the boresight target offset for the appropriate aided vision device. For laser aiming lights, position the weapon so the bore light strikes the black dot on the boresight target.

(a) Adjust the *laser* until it centers on the cross on the boresight target offset. For aided vision optics, position the weapon so the *reticle* centers on the cross on the boresight target offset (the firer must physically aim the weapon).

(b) Adjust the aided vision optic until the bore light strikes the black dot on the boresight target. Refer to specific instructions on each boresight-offset target.

(c) You have boresighted the weapon when both the laser bore light and the aiming point or laser move to the appropriate position on the boresight target offset.

G-11. TARGET OFFSETS

Figure G-12 (page G-23), Figure G-13 (page G-24), and Figure G-14 (page G-25) show example boresight target offsets. The gunner measures boresight offsets in 1-centimeter squares. For targets, he copies M16A2 25-meter zero targets, with the silhouette representing the point of aim.

a. **Aligning the M2 with the AN/PEQ-2A**. Using the 10-meter boresight target, align the bore light on the dot and adjust the AN/PEQ-2A to the cross. This gives an 800-meter zero (Figure G-12, page G-23).

b. **Aligning the M2 with the AN/TVS-5**. Using the 10-meter boresight target, place the target at a distance of 10 meters. Aim the 400-meter AN/TVS-5 reticle at the cross and adjust the sight so that the bore light strikes the dot (Figure G-13, page G-24).

c. **Aligning the M2 with the AN/PAS-13**. Using the 10-meter boresight target, place the tips of the index fingers on the gray circles. Aim between the hot spots provided by the fingers. Adjust the TWS so that the bore light strikes the dot when you aim between the hot spots at 10 meters. If using a spacer, increase the offset by the height of the spacer (Figure G-14, page G-25).

10-meter boresight target AN/PEQ-2A mounted on an AN/TVS-5 feed tray cover bracket and plastic spacer on M2 MG

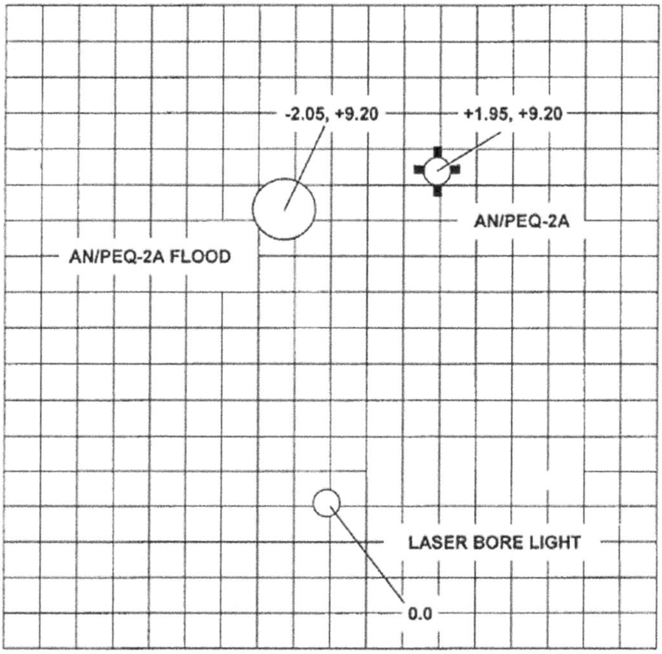

This is the 10-meter boresight target for 500-meter zero.
Grids are 1-centimeter square.
Units manufacture targets locally.

Zero target data for M2 MG:

1. **Stabilize the weapon.**
2. **Align laser borelight on its dot.**
3. **Adjust the AN/PEQ-2A until aiming laser centers on the dot cross hair.**

Figure G-12. Alignment of M2 with the AN/PEQ-2A.

10-meter boresight target
AN/TVS-5 mounted on AN/TVS-5 feed tray cover bracket on the M2 MG.

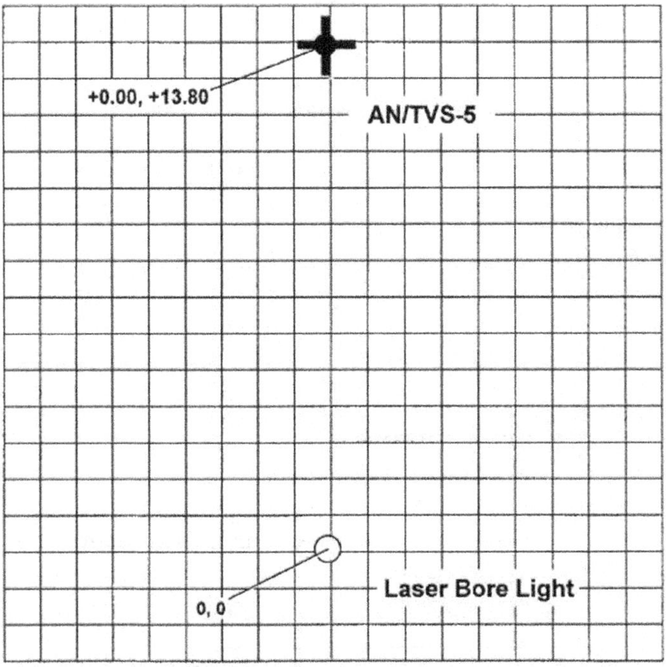

This is the 10-meter boresight target for the 500-meter zero. Grids measure 1 cm square. Units manufacture targets locally.

Zero target data for M2 MG:

1. Stabilize the weapon.
2. Aim the AN/TVS-5 at the cross hair and adjust until the borelight strikes the laser's borelight dot.

Figure G-13. Alignment of M2 with the AN/TVS-5.

10-meter boresight target AN/PAS-13 mounted on AN/PAS-13 feed tray cover bracket on the M2 MG.

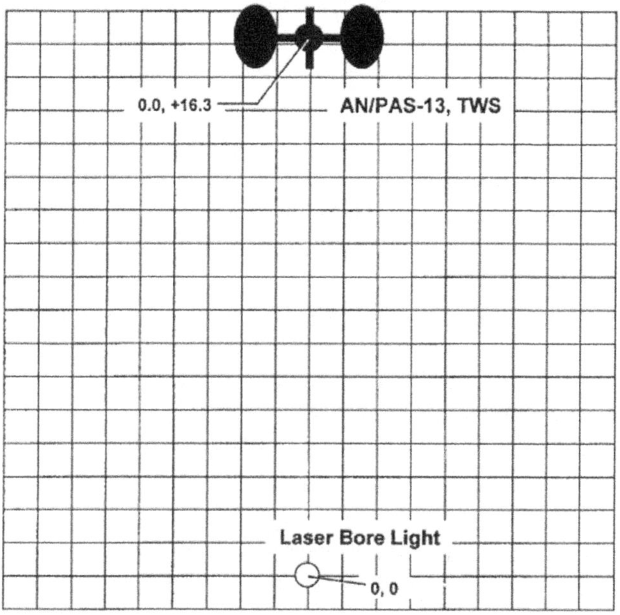

This is the 10-meter boresight target for the 500-meter zero. Grids measure 1 cm square. Units manufacture targets locally.

Zero target data for M2 MG:

1. Stabilize the weapon.
2. Place the thermal signature material or the tips of fingers on the gray circles.
3. Aim between the gray circles at the cross hairs.
4. Adjust the AN/PAS-13 until the borelight laser centers on its dot.
5. Zero the reticle in *both* wide and narrow fields of view.

Figure G-14. Alignment of M2 with the AN/PAS-13.

Section V. T&E MANIPULATION

The T&E mechanism allows engagement of preselected target areas at night or during limited visibility. The gunner records (in mils) the direction and elevation readings from the traversing bar and T&E mechanism.

G-12. ZEROING THE T&E MECHANISM

Before a gunner can effectively engage targets with the T&E, he must zero it to the weapon:

a. **Zero Traversing Handwheel.** Hold the T&E so the traversing handwheel sits on the left as you look at it.

(1) Turn the traversing handwheel toward you until it stops.

(2) Loosen the locking nut slightly.

(3) Align the zero on the scale with the zero on the elevating screw yoke.

(4) Hold the scale (with the zeroes aligned) and tighten the locking nut. Make sure the zeroes remain aligned.

(5) Turn the traversing handwheel two complete revolutions away from you. If doing this at night, count 50 "clicks" away from you.

b. **Zero Elevating Handwheel to Upper Elevating Screw.** Align the two zeroes.

(1) Rotate the elevating handwheel up or down until you can see a zero with a line below it on the upper elevating screw.

(2) Position the elevating handwheel so the indicator points to the zero on the handwheel.

c. **Zero Elevating Mechanism Sleeve to Lower Elevating Screw.**

(1) Rotate the elevating mechanism sleeve all the way up, and then rotate it down until it stops. Note the number of complete turns down.

(2) Rotate the elevating mechanism sleeve up half that many turns.

(3) Position the slide lock lever to face you.

(4) Attach the T&E to the tripod and gun.

G-13. LAYING THE GUN FOR DIRECTION

After the gunner receives an assigned sector of fire, he should--

a. Pick up the rear legs of the tripod.

b. Shift the tripod until the muzzle of the weapon points to the center of the sector of fire.

c. After laying the weapon for direction, firmly stamp in the tripod shoes and place sandbags on the legs. This aids stability and may prevent accidental movement.

d. Obtain and record directional readings to all targets within the sector and perform the following:

(1) Loosen the traversing slide lock lever and move the slide along the traversing bar until the weapon lays either on the center of a point target or on the flank of a linear target.

(2) Lock the traversing bar. Read the direction from the scale on the traversing bar. If the left edge of the traversing slide falls anywhere

outside the 5-mil tick mark, the gunner moves the left edge of the traversing bar slide back to the next smaller mil reading. Then, the gunner uses the traversing handwheel to complete the initial lay.

e. Obtain a reading to the target by the direction of the weapon barrel. If the barrel is moved to the right, the gunner records a right heading. He reads the number on the traversing bar from the left side of the traversing slide lock. If the barrel is moved to the left and the traversing slide lock is on the right side of the zero, he records a left reading.

f. After taking a directional reading for a target, measure the width of the target in mils. Then, he uses the traversing handwheel to move the barrel across the target--one click equals one mil.

g. Before moving to another target, reposition the traversing mechanism.

h. Obtain elevation readings. The gunner verifies that the weapon is laid on the center base of the target. Read the elevation from two scales.

i. Get the first, or major, part of the elevation from the elevating screw plate scale.

j. Get the second, or minor, part of the elevation from the elevating handwheel.

k. Separate the two parts of the elevation reading with a slash (/). For example, the gunner could write "-50/3." An elevation reading might apply only to the T&E where he reads it.

l. Note that, if the number of threads increases or decreases after he records the data, he cannot place accurate fire on the target. For example, rotating the base of the T&E mechanism to engage a secondary sector makes the data incorrect--unless the same number of threads appear both before and after the move.

* GLOSSARY

AP	armor piercing
APC	armored personnel carrier
approx	approximately
attn	attention
ba	battery
BFA	blank firing attachment
BFV	Bradley fighting vehicle
BMP	Soviet fighting vehicle
cal	caliber
cdr	commander
CLP	cleaner, lubricant, and preservative
CVKI	combat vehicle kill indicator
DA	Department of the Army
FEBA	forward edge of the battle area
FM	field manual
FOV	field of view
FPF	final protective fires
FPL	final protective line
FSN	federal stock number
FTX	field training exercise
GA	Georgia
HB	heavy barrel
HMMWV	high-mobility, multi-purpose, wheeled vehicle
HWTS	heavy weapon thermal sight
IAW	in accordance with
IN	infantry
IR	infrared
LED	light-emitting diode
LFX	live-fire exercise
LSA	semifluid lubricating oil
LTA	local training area
m	meter
METL	mission-essential task list
MG	machine gun
MILES	multiple integrated laser engagement system
mm	millimeter
MOPP	mission-oriented protection posture
mph	miles per hour
MOUT	military operations on urbanized terrain
MTA	major training area
MTP	mission training plan
NBC	nuclear, biological, and chemical
NCO	noncommissioned officer

NCOES	noncommissioned officer education system
NCOIC	noncommissioned officer in charge
NSN	national stock number
NVD	night vision device
OIC	officer in charge
PDF	principal direction of fire
PL-M	lubricating oil, general purpose
PL-S	lubricating oil, special purpose
POI	program of instruction
RABA	recoil amplifier barrel assembly
RBC	rifle bore cleaner
SM	soldiers manual
SMCT	soldiers manual of common tasks
SOP	standing operating procedure
SRTA	short-range training ammunition
SRTA-T	short-range training ammunition-tracer
STP	soldiers training publication
STRAC	Standards in Training Commission
STX	situational training exercise
T&E	traversing and elevating
TM	technical manual
TRADOC	Training and Doctrine Command
TRP	target reference point
US	United States
v	volt

* REFERENCES

SOURCES USED

These are the Army publications quoted or paraphrased in this manual.

DA Pam 350-38	Standards in Weapons Training. 3 July 1997
FM 3-5	NBC Decontamination. 17 November 1993.
FM 44-80	Visual Aircraft Recognition. 30 September 1996.
STP 21-24-SMCT	Soldier's Manual of Common Tasks, Skill Levels 2/3/4. 1 October 1992.
TC 25-8	Training Ranges. 25 February 1992, with Change 1, 16 November 1992.
TM 9-1005-213-10	Operator's Manual for Machine Guns, Caliber .50; Browning, M2, Heavy Barrel Flexible, 6 August 1992, with Changes 1 and 2, 1 August 1995 and 6 August 1996.
TM 9-1300-200	Ammunition General. 3 October 1969; with changes 1 through 5, dated 22 January 1971 through 30 September 1993.
TM 11-5855-214-10	Operator's Manual for Night Vision Sight, Crew Served Weapon, AN/TVS-5. 15 February 1989.
TM 11-5855-302-12&P	Operator's and Unit Maintenance Manual (including Repair Parts and Special Tools List), Sight, Thermal, AN/PAS-13(V)2, AN/PAS-13(V)3. To be published.
TM 11-5855-312-10	Operator's Manual, Sight, Thermal, AN/PAS-13B(V)2, AN/PAS-13B(V)3. To be published.

DOCUMENTS NEEDED

These Army publications must be available to the intended users of this manual.

Army Forms

These forms can be locally reproduced on 81/2 x 11-inch paper. The reproducible form is located at the back of this manual. The forms are also available on the Army Electronic Library (AEL) CD-ROM (EM0001) and at the USAPA website located at (www.usapa.army.mil):

DA Form 2404	Equipment Inspection and Maintenance Worksheet. 1 April 1979.
DA Form 5517-R	Standard Range Cards (LRA). February 1986.
DA Form 7007-R	Machine Gun Scorecard for M2. August 2001.
DA Form 7448-R	M2 Caliber .50 Heavy Barrel Machine Gun, Firing Table I, Day Practice Scorecard. January 2002.
DA Form 7449-R	M2 Caliber .50 Heavy Barrel Machine Gun, Firing Table II, Day Qualification Scorecard. January 2002.
* DA Form 7450-R	M2 Caliber .50 Heavy Barrel Machine Gun, Firing Tables III(A) and III(B), Night Practice Scorecard. October 2002.
* DA Form 7451-R	M2 Caliber .50 Heavy Barrel Machine Gun, Firing Tables IV(A), Night Qualification Scorecard; and IV(B), Day Practice Scorecard. October 2002.

Technical Manuals

TM 11-5855-306-10	Operator's Manual Monocular Night Vision Device (MNVD), AN/PVS-14. June 2000.
TM 11-5855-306-23&P	Unit and Direct Support Maintenance Manual (Including Repair Parts and Special Tools List), Monocular Night Vision Device (MNVD), AN/PVS-14. June 2000.
TM 11-5855-309-12&P	Operator's and Unit Maintenance Manual (including Repair Parts and Special Tools List), Sight, Thermal, AN/PAS-13A(V)2, AN/PAS-13A(V)3. To be published.

INTERNET WEB SITES

U.S. Army Publishing Agency (USAPA):
www.usapa.army.mil

Reimer Digital Library (RDL):
www.adtdl.army.mil

Air Force Publishing:
afpubs.hq.af.mil/

* INDEX

MACHINE GUN SCORECARD FOR M2

For use of this form, see FM 23-65, Change 1, or the back of this form. The proponent agency is TRADOC.

DATA REQUIRED BY THE PRIVACY ACT OF 1974

AUTHORITY: 10 USC 3012(g)/Executive order 9397.

PRINCIPAL PURPOSE: To aid individual training on targets at various ranges.

ROUTINE USE: To evaluate individual proficiency. SSN is used for positive identification purposes only.

DISCLOSURE: Voluntary. However, mass rating and scoring require some tracking method. Therefore, only those soldiers who provide their SSNs can receive scores or qualify on weapons or equipment.

1a. LAST NAME	1b. FIRST NAME	1c. MI	2. DATE *(YYYYMMDD)*

3. SSN	4. UNIT	5. LANE

	6a. FIRING TABLE I			7a. FIRING TABLE II					8a. FIRING TABLE III					9a. FIRING TABLE IV				
TSK	RANGE (M)	HIT	PTS	RANGE (M)	TIME	HIT	PTS	BON	RANGE (M)	TIME	HIT	PTS	BON	RANGE (M)	TIME	HIT	PTS	BON
1	10	NA	NA	550	None	NA	NA	NA	550	None	NA	NA	NA	550	None	NA	NA	NA
2	10	NA	NA	800	20 Sec				800	20 Sec				800	20 Sec			
3	10	NA	NA	400	20 Sec				400	20 Sec				400	20 Sec			
4	10	NA	NA	700	25 Sec				700	25 Sec				700	25 Sec			
5	10			1,000	25 Sec				1,000	25 Sec				1,000	25 Sec			
6	10			400	35 Sec				400	35 Sec				400	35 Sec			
				700					700					700				
7	10			550	35 Sec				550	35 Sec				550	35 Sec			
				800					800					800				
8	NA	NA	NA	400	45 Sec				400	45 Sec				400	45 Sec			
				550					550					550				
				1,000					1,000					1,000				

6b. SUBTOTAL FIRING TABLE I	7b. SUBTOTAL FIRING TABLE II	8b. SUBTOTAL FIRING TABLE III	9b. SUBTOTAL FIRING TABLE IV

10. TOTAL SCORE	11. RATING
12a. OIC's PRINTED OR TYPED NAME	13a. GRADER'S PRINTED OR TYPED NAME
12b. OIC'S SIGNATURE	13b. GRADER'S SIGNATURE

COMPLETE THE BLOCKS ON THIS SCORECARD AS FOLLOWS:

1. NAME. Enter last name, first name, middle initial.

2. DATE. Enter today's date in YYYYMMDD format.

3. SSN. Enter gunner's social security number.

4. UNIT. Enter gunner's unit.

5. LANE. Enter gunner's firing lane.

6a - 9a. FIRING TABLE I thru IV

HIT column. As applicable, mark block with an "X" if gunner hit target.

POINTS column. As applicable, enter points earned for each target hit.

BON *(Bonus)* column *(Firing Tables I thru III only)*. As applicable, enter bonus points earned for each target hit.

6b-9b. SUBTOTALS. Subtotal all of the points earned for each firing table.

10. TOTAL SCORE. Add firing table subtotals *(Blocks 6b - 9b)* to obtain this score.

11. RATING. Compare total score *(Block 10)* to rating chart to determine the soldier's rating.

12a. OIC's PRINTED OR TYPED NAME. Print or type OIC's name.
12b. OIC's SIGNATURE. *(OIC must sign the actual form).*

13a. GRADER's PRINTED OR TYPED NAME. Print or type grader's name.
13b. GRADER's SIGNATURE. *(Grader must sign the actual form).*

COMPUTE RATING:

Maximum..............................	218 points
Expert..................................	196 to 217 points
Gunner first class......................	174 to 195 points
Gunner second class................	153 to 173 points
Unqualified..............................	152 points or less

DA FORM 7007-R, JAN 2002 (Back) USAPA V1.00

M2 CALIBER .50 HEAVY BARREL MACHINE GUN FIRING TABLE I
DAY PRACTICE SCORECARD

For use of this form, see FM 23-65, Change 1; the proponent agency is TRADOC.

DATA REQUIRED BY THE PRIVACY ACT OF 1974

AUTHORITY: 10 USC 3012(g)/Executive order 9397.
PRINCIPAL PURPOSE: To aid individual training on targets at various ranges.
ROUTINE USES: To evaluate individual proficiency. SSN is used for positive identification purposes only.
DISCLOSURE: Voluntary. However, mass rating and scoring require some tracking method. Therefore, only those soldiers who provide their SSNs can receive scores or qualify on weapons or equipment.

1a. LAST NAME	1b. FIRST NAME	1c. MI	2. DATE *(YYYYMMDD)*

3. SSN	4. UNIT	5. RANGE	6. LANE

7a. ENGAGEMENT	7b. STANDARDS	7c. AMMO	7d. TIME *(Minutes)*	7e. RANGE *(Meters)*	7f. GO	7g. NO GO
ZERO	1 BURST HIT	14	NA	400	NA	NA
2	1 BURST HIT	14	1.5	1,100 (±200)		
3	1 BURST HIT	14	1.5	1,500 (±200)		
4	1 BURST HIT	14	1.5	600 (±100)		
5	1 BURST HIT	14	1.5	800 (±100)		
6	1 BURST HIT	14	1.5	400		
8. MULTIPLE TARGETS						
7	1 BURST HIT	28	2.5	1,100 (±200)		
	1 BURST HIT			600 (±100)		
8	1 BURST HIT	28	2.5	800 (±100)		
	1 BURST HIT			1,500 (±200)		

9. SCORING *(Check one)*

EXPERT - 9 ☐ SHARPSHOOTER - 8 ☐

MARKSMAN - 6 ☐ UNQUALIFIED - 5 OR BELOW ☐

10. GUNNER'S SIGNATURE

11a. GRADER'S PRINTED OR TYPED NAME	11b. GRADER'S SIGNATURE

M2 CALIBER .50 HEAVY BARREL MACHINE GUN FIRING TABLE II
DAY QUALIFICATION SCORECARD

For use of this form, see FM 23-65, Change 1; the proponent agency is TRADOC.

DATA REQUIRED BY THE PRIVACY ACT OF 1974

AUTHORITY: 10 USC 3012(g)/Executive order 9397.

PRINCIPAL PURPOSE: To aid individual training on targets at various ranges.

ROUTINE USES: To evaluate individual proficiency. SSN is used for positive identification purposes only.

DISCLOSURE: Voluntary. However, mass rating and scoring require some tracking method. Therefore, only those soldiers who provide their SSNs can receive scores or qualify on weapons or equipment.

1a. LAST NAME	1b. FIRST NAME	1c. MI	2. DATE *(YYYYMMDD)*

3. SSN	4. UNIT	5. RANGE	6. LANE

7a. ENGAGEMENT	7b. STANDARDS	7c. AMMO	7d. TIME *(Minutes)*	7e. RANGE *(Meters)*	7f. GO	7g. NO GO
ZERO	1 BURST HIT	14	NA	400	NA	NA
2	1 BURST HIT	14	1	1,100 (±200)		
3	1 BURST HIT	14	1	1,500 (±200)		
4	1 BURST HIT	14	1	600 (±100)		
5	1 BURST HIT	14	1	800 (±100)		
6	1 BURST HIT	14	1	400		
8. MULTIPLE TARGETS						
7	1 BURST HIT	28	2	1,100 (±200)		
	1 BURST HIT			600 (±100)		
8	1 BURST HIT	28	2	800 (±100)		
	1 BURST HIT			1,500 (±200)		

9. SCORING *(Check one)*

EXPERT - 9 ☐ SHARPSHOOTER - 8 ☐

MARKSMAN - 6 ☐ UNQUALIFIED - 5 OR BELOW ☐

10. GUNNER's SIGNATURE

11a. GRADER'S PRINTED OR TYPED NAME	11b. GRADER'S SIGNATURE

DA FORM 7449-R, JAN 2002

USAPA V1.00

M2 CALIBER .50 HEAVY BARREL MACHINE GUN FIRING TABLES III (A) AND III (B) NIGHT PRACTICE SCORECARDS

For use of this form, see FM 23-65, Change 2; the proponent agency is TRADOC.

DATA REQUIRED BY THE PRIVACY ACT OF 1974

AUTHORITY: 10 USC 3012(g)/Executive order 9397.
PRINCIPAL PURPOSE: To aid individual training on targets at various ranges.
ROUTINE USES: To evaluate individual proficiency. SSN is used for positive identification purposes only.
DISCLOSURE: Voluntary. However, mass rating and scoring require some tracking method. Therefore, only those soldiers who provide their SSNs can receive scores or qualify on weapons or equipment.

1a. LAST NAME	1b. FIRST NAME	1c. MI	1d. RANK	2. SSN

TABLE III (A). NIGHT PRACTICE SCORECARD

NOTE: Use this table if the M2 has the TWS, or if it has an AN/TVS-5 that has been upgraded with the third-generation tube.

a. RANGE	b. UNIT	c. LANE	d. DATE *(YYYYMMDD)*

e. ENGAGEMENT	f. STANDARDS	g. AMMO	h. TIME *(Minutes)*	i. RANGE *(Meters)*	j. GO	k. NO GO
ZERO	1 BURST HIT	14	NA	400	NA	NA
2	1 BURST HIT	14	1.5	1,100 (±200)		
3	1 BURST HIT	14	1.5	1,500 (±200)		
4	1 BURST HIT	14	1.5	600 (±100)		
5	1 BURST HIT	14	1.5	800 (±100)		
6	1 BURST HIT	14	1.5	400		
l. MULTIPLE TARGETS						
7	1 BURST HIT	28	2.5	1,100 (±200)		
	1 BURST HIT			600 (±100)		
8	1 BURST HIT	28	2.5	800 (±100)		
	1 BURST HIT			1,500 (±200)		

4. SCORING *(Check one)*
- EXPERT - 9 ☐
- SHARPSHOOTER - 8 ☐
- MARKSMAN - 6 ☐
- UNQUALIFIED - 5 OR BELOW ☐

5. GUNNER'S SIGNATURE

6a. GRADER'S PRINTED OR TYPED NAME

6b. GRADER'S SIGNATURE

7. TABLE III (B). NIGHT PRACTICE SCORECARD

NOTE: Use this table if the M2 has an AN/TVS-5 that has not been upgraded.

a. RANGE	b. UNIT	c. LANE	d. DATE *(YYYYMMDD)*

e. ENGAGEMENT	f. STANDARDS	g. AMMO	h. TIME *(Minutes)*	i. RANGE *(Meters)*	j. GO	k. NO GO
ZERO	1 BURST HIT	14	NA	400	NA	NA
2	1 BURST HIT	14	1.5	1,100 (±200)		
3	1 BURST HIT	14	1.5	800 (±100)		
4	1 BURST HIT	14	1.5	600 (±100)		
5	1 BURST HIT	14	1.5	800 (±100)		
6	1 BURST HIT	14	1.5	400		
l. MULTIPLE TARGETS						
7	1 BURST HIT	28	2.5	1,100 (±200)		
	1 BURST HIT			600 (±100)		
8	1 BURST HIT	28	2.5	800 (±100)		
	1 BURST HIT			600 (±100)		

8. SCORING *(Check one)*
- EXPERT - 9 ☐
- SHARPSHOOTER - 8 ☐
- MARKSMAN - 6 ☐
- UNQUALIFIED - 5 OR BELOW ☐

9. GUNNER'S SIGNATURE

10a. GRADER'S PRINTED OR TYPED NAME

10b. GRADER'S SIGNATURE

M2 CALIBER .50 HEAVY BARREL MACHINE GUN FIRING TABLES IV (A) AND IV (B) NIGHT QUALIFICATION SCORECARDS

For use of this form, see FM 23-65, Change 2; the proponent agency is TRADOC.

DATA REQUIRED BY THE PRIVACY ACT OF 1974

AUTHORITY: 10 USC 3012(g)/Executive order 9397.
PRINCIPAL PURPOSE: To aid individual training on targets at various ranges.
ROUTINE USES: To evaluate individual proficiency. SSN is used for positive identification purposes only.
DISCLOSURE: Voluntary. However, mass rating and scoring require some tracking method. Therefore, only those soldiers who provide their SSNs can receive scores or qualify on weapons or equipment.

1a. LAST NAME	1b. FIRST NAME	1c. MI	1d. RANK	2. SSN

3. TABLE IV (A). NIGHT QUALIFICATION SCORECARD

NOTE: Use this table if the M2 has the TWS, or if it has an AN/TVS-5 that has been upgraded with the third-generation tube.

a. RANGE	b. UNIT	c. LANE	d. DATE *(YYYYMMDD)*

e. ENGAGEMENT	f. STANDARDS	g. AMMO	h. TIME *(Minutes)*	i. RANGE *(Meters)*	j. GO	k. NO GO
ZERO	1 BURST HIT	14	NA	400	NA	NA
2	1 BURST HIT	14	1	1,100 (±200)		
3	1 BURST HIT	14	1	1,500 (±200)		
4	1 BURST HIT	14	1	600 (±100)		
5	1 BURST HIT	14	1	800 (±100)		
6	1 BURST HIT	14	1	400		
l. MULTIPLE TARGETS						
7	1 BURST HIT	28	2	1,100 (±200)		
	1 BURST HIT			600 (±100)		
8	1 BURST HIT	28	2	800 (±100)		
	1 BURST HIT			1,500 (±200)		

4. SCORING *(Check one)*
EXPERT - 9 ☐
SHARPSHOOTER - 8 ☐
MARKSMAN - 6 ☐
UNQUALIFIED - 5 OR BELOW ☐

5. GUNNER'S SIGNATURE

6a. GRADER'S PRINTED OR TYPED NAME

6b. GRADER'S SIGNATURE

7. TABLE IV (B). NIGHT QUALIFICATION SCORECARD

NOTE: Use this table if the M2 has an AN/TVS-5 that has not been upgraded.

a. RANGE	b. UNIT	c. LANE	d. DATE *(YYYYMMDD)*

e. ENGAGEMENT	f. STANDARDS	g. AMMO	h. TIME *(Minutes)*	i. RANGE *(Meters)*	j. GO	k. NO GO
ZERO	1 BURST HIT	14	NA	400	NA	NA
2	1 BURST HIT	14	1	1,100 (±200)		
3	1 BURST HIT	14	1	800 (±100)		
4	1 BURST HIT	14	1	600 (±100)		
5	1 BURST HIT	14	1	800 (±100)		
6	1 BURST HIT	14	1	400		
l. MULTIPLE TARGETS						
7	1 BURST HIT	28	2	1,100 (±200)		
	1 BURST HIT			600 (±100)		
8	1 BURST HIT	28	2	800 (±100)		
	1 BURST HIT			600 (±100)		

8. SCORING *(Check one)*
EXPERT - 9 ☐
SHARPSHOOTER - 8 ☐
MARKSMAN - 6 ☐
UNQUALIFIED - 5 OR BELOW ☐

9. GUNNER'S SIGNATURE

10a. GRADER'S PRINTED OR TYPED NAME

10b. GRADER'S SIGNATURE

FM 3-22.31 (FM 23-31)
C1

Change 1
Field Manual
No. 3-22.31

Headquarters
Department of the Army
Washington, DC, 19 March 2007

40-mm Grenade Launcher, M203

1. Change FM 3-22.31, 13 February, 2003, as follows:

Remove old pages:	**Insert new pages:**
i through iv	I through iv
1-1 through 1-2	1-1 through 1-2
Chapter 2	Chapter 2
Chapter 3	Chapter 3
Chapter 4	Chapter 4
Chapter 5	Chapter 5
6-9 through 6-10	6-9 through 6-10
Glossary	Glossary
References	References
Index	Index
DA Form 2946-R	DA Form 2946-R

2. A star (*) marks new or changed material.

3. File this transmittal sheet in front of the publication.

DISTRIBUTION RESTRICTION: Approved for public release; distribution is unlimited.

By Order of the Secretary of the Army:

PETER J. SCHOOMAKER
General, United States Army
Chief of Staff

Official:

Joyce E. Morrow

JOYCE E. MORROW
Administrative Assistant to the
Secretary of the Army
0705702

DISTRIBUTION:
Regular Army, USAR, and ARNG: To be distributed in accordance with initial distribution number 115893, requirements for FM 3-22.31.

This page intentionally left blank.

FIELD MANUAL
NO. 3-22.31 (23-31)

HEADQUARTERS
DEPARTMENT OF THE ARMY
WASHINGTON, D.C., 19 March 2007

40-MM GRENADE LAUNCHER, M203

CONTENTS

DISTRIBUTION RESTRICTION: Approved for public release; distribution is unlimited.

*This publication supersedes FM 23-31, 20 September 1994.

PREFACE

This publication provides technical information on and training and combat techniques for the M203 grenade launcher. Intended users include leaders and designated grenadiers, who will use this information to successfully integrate the M203 into their combat operations. This manual discusses gunnery training and train-the-trainer, and it includes an appendix on the 40-mm grenade launcher, M79.

*This publication applies to the Regular Army, the Army National Guard (ARNG)/Army National Guard of the United States (ARNGUS), and the United States Army Reserve (USAR) unless otherwise stated.

*The tactical positions shown in this manual were drawn to enhance the reader's understanding of related subject material and are *not* tactically correct. Also, to avoid hiding important graphic information, uniforms may be shown plain rather than camouflaged.

*This publication prescribes DA Form 2946-R, *40-MM Grenade Launcher Scorecard.*

*The proponent for this publication is the US Army Training and Doctrine Command. The preparing agency is the US Army Infantry School. You may send comments and recommendations by any means, US mail, e-mail, fax, or telephone, as long as you use or follow the format of DA Form 2028, *Recommended Changes to Publications and Blank Forms*. You may also phone for more information:

E-mail lu.lusano@us.army.mil
Phone COM (706) 545-8623 or DSN 835-8623
Fax COM (706) 545-8600 or DSN 835-8600
US Mail Cmdt, USAIS
ATTN: ATSH-INB
6650 Wilkin Drive , Bldg 74, Rm 102
Fort Benning, GA 31905-5593

*Unless this publication states otherwise, masculine nouns and pronouns may refer to either men or women.

CHAPTER 1
INTRODUCTION

This chapter discusses the training strategy and combat conditions for the 40-mm grenade launcher, M203. (Appendix A discusses the M79 model.)

* 1-1. TRAINING STRATEGY

An effective overall training strategy produces well-trained grenadiers and trainers by integrating resources into an effective year-round training program. Beginning with IET and continuing both in other institutions (NCOES, OBC, and CCC) and in the unit, such a program trains and sustains the individual and collective skills needed to perform the wartime mission. Specific training strategies are implemented by institutional and unit training programs, and supporting training strategies are implemented through the use of other resources such as publications, ranges, ammunition, training aids, devices, simulators, and simulations. The year-round program includes periodic preliminary marksmanship training followed by zeroing and range qualification firing. Other key elements of the program are training for the trainers and refresher training for nonfiring skills. The example in Figure 1-1 on page 1-3 shows the flow of unit sustainment training.

*a. **Institutional Training**. Training strategy begins with combat arms initial-entry training (IET), which trains soldiers in the standards of M203 gunnery tasks. Soldiers graduate with basic and advanced M203 skills that include maintaining the M203 and using it to hit a variety of targets. Other institutional training programs, such as NCOES, OBC, and CCC, reinforce these skills. Related soldier skills are integrated into tactical training (STP 21-1-SMCT).

b. **Unit Training**. Training continues in units, where, in addition to sustaining proficiency in skills gained in institutional training, leaders and soldiers develop and sustain new skills such as suppressive and supporting fire. These skills are integrated into collective training exercises to develop combat readiness. Preliminary marksmanship training is conducted before firings and as other opportunities arise. (Appendix B discusses an M203 unit training program.) To be effective, a unit training program focuses on three battlefield variables:

(1) ***Target***. Is the target moving or stationary, single or multiple?

(2) ***Grenadier***. Is the grenadier moving or stationary? Is he kneeling, prone, or standing?

(3) ***Conditions***. Is visibility full or limited? Must soldiers wear protective masks or not? Is it day or night?

c. **Initial and Sustainment Training**. A task that is taught correctly and learned well is retained longer, so initial training is critical. In addition to being more easily sustained, well-trained skills are also easier to regain if not used for some time. Retraining may be needed, however, if too much time elapses, if training doctrine changes, or if personnel turnover is high.

(1) ***Collective Training***. Collective training exercises progress from drills (squad, section, and platoon) to STXs, and then to live-fire tactical exercises (LFXs). Drill books

and MTPs provide tasks and guidance needed to plan and conduct the exercises. After each, leaders and trainers conduct an AAR to evaluate both individual and unit proficiency. The results provide readiness indicators and requirements for future training. LFXs provide leaders with an overview of unit proficiency and training effectiveness. They can be conducted on any range approved for M203 firing.

(2) ***Leader Training***. The most critical part of the Army's overall gunnery training strategy is to train the trainers and leaders first. Leader courses, however, include only limited M203 training, so officers and NCOs should use available publications to develop their proficiency with the M203. Publications help leaders plan, conduct, and evaluate their gunnery training programs. Proponent schools provide training support materials (field manuals, training aids, devices, simulators, and audiovisual programs), which provide the doctrinal foundations for training the force.

(3) ***Advanced Training***. Once the soldier knows the weapon and has demonstrated skill in zeroing, training strategy provides for additional live-fire training and target-acquisition exercises, which are conducted at various ranges. To develop proficiency, soldiers must master different types of targets and scenarios of increasing difficulty.

(4) ***Proficiency Assessment***. This is conducted on the zeroing and record live-fire exercise range when soldiers complete IET.

1-2. COMBAT CONDITIONS

The trainer must realize that qualification is not an end but a step toward reaching combat readiness. To reach combat readiness, the grenadier should consider his position, the capabilities of his weapon, and the following combat conditions:

a. Enemy personnel are seldom visible except when assaulting.

b. Most combat fire must be directed at an area where the enemy has been detected or is suspected but cannot be seen. Area targets consist of objects or outlines of men irregularly spaced along covered and concealed areas (ground folds, hedges, borders of woods).

c. Most combat targets can be detected by smoke, flash, dust, noise, or movement and are visible only for a moment.

d. Some combat targets can be engaged by using reference points, predetermined fire, or range card data.

e. The nature of the target and irregularities of terrain and vegetation may require a grenadier to use a variety of positions to place effective fire on the target. In a defensive situation, the grenadier usually fires from a supported fighting position.

f. Most combat targets have a low-contrast outline and are obscured. Therefore, choosing an aiming point in elevation is difficult.

g. Time-stressed fire in combat can be divided into three types:

- A single, fleeing target that must be engaged quickly.
- Area targets engaged within the time they remain available.
- A surprise target that must be engaged at once with instinctive, accurate fire.

*CHAPTER 2

DESCRIPTION AND MAINTENANCE

Proper weapon maintenance is a vital part of all gunnery training programs. Good maintenance contributes to weapon effectiveness as well as to unit readiness. This chapter provides a technical description of the M203 grenade launcher, its components, and its ammunition. It also discusses proper procedures for clearing, disassembling, cleaning, lubricating, inspecting, and caring for the weapon.

2-1. DESCRIPTION

The M203-series grenade launchers are lightweight, single-shot, breech-loaded, pump action (sliding barrel), shoulder-fired weapons (Figure 2-1).

a. The original model of the M203 was designed for use with M16-series rifles.

b. The M203A1 model was designed for use with M4 series carbines.

c. The M203A2 model has a quick-release bracket for use with M4 and M4A1 carbines that have the M4 rail system and with M16A4 rifles that have the M5 adapter rail system.

Model	Who Mounts/Dismounts	Weapon Mounted On
M203	Unit armorer	M16-series rifles
M203A1/A2	DS maintenance co	M4-series carbine rifles

NOTE: Individual soldiers should neither mount nor dismount the grenade launcher.

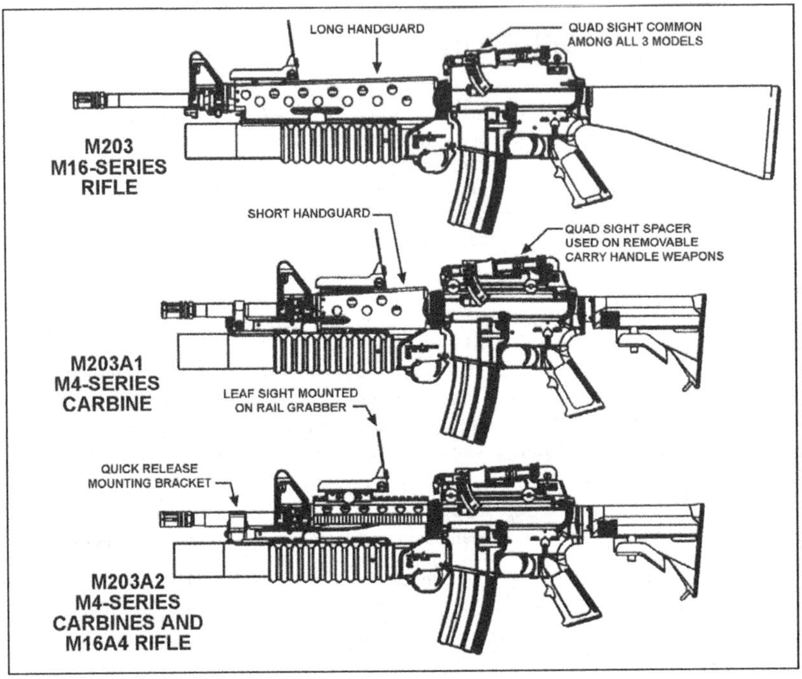

Figure 2-1. M203-series grenade launchers.

2-2. TECHNICAL DATA

Technical data for the M203-series grenade launchers follow:

a. **Weapon.**

Length
- Rifle and grenade launcher (overall) .... 99.0 cm (39 inches)
- Barrel only .... 30.5 cm (12 inches)
- Rifling .... 25.4 cm (10 inches)

Weight
- Launcher, unloaded .... 1.4 kg (3.0 pounds)
- Launcher, loaded .... 1.6 kg (3.5 pounds)
- Rifle and grenade launcher, both fully loaded .... 5.0 kg (11.0 pounds)

Number of lands .... 6 right hand twist

b. **Ammunition.**

Caliber .... 40 mm
Weight .... About 227 g (8 oz)

c. **Operational Characteristics.**

Action .... Single shot

Sights (use together or separately)

Front leaf sight	Use between 50 and 250 meters (beyond 250 meters, the M16's barrel hides the target)
Rear quadrant sight	Use with M16's front sight post for ranges 50 to 250 meters.
Integrated front sight post, rear sight aperture	Use for 50- to 400 meters.
Chamber pressure	82,737 kilopascals (12,000 psi)
Muzzle velocity	76 mps (250 fps)
Maximum range	About 400 m (1,320 ft)
Maximum effective range	
Fire-team sized area target	350 m (1,155 ft)
Vehicle or weapon point target	150 m (495 ft)
Minimum safe firing range (HE)	
Training unprotected	165 m (541 ft)
Protected firing position	Less than 165 m (541 ft)
Combat	31 m (102 ft)
Minimum arming range	14 to 38 m (46 to 125 ft)
Rate of fire	5 to 7 rounds per minute
Minimum combat load	36 HE rounds

WARNING

When firing close-in, such as in urban areas, trenches, and other restrictive terrain, consider the minimum arming range to ensure the round clears other friendly forces.

2-3. COMPONENTS

Figure 2-2 shows the M203's major components, and the following paragraphs describe their purposes. Figure 2-3 through Figure 2-6 show the sight assemblies, the trigger and trigger guard, and the safety:

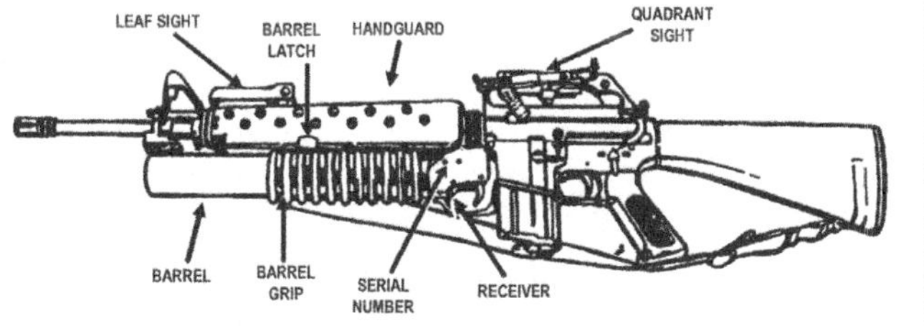

Figure 2-2. Components of the M203 grenade launcher.

a. **Handguard**. The handguard assembly houses the rifle barrel (Figure 2-2).

b. **Quadrant Sight Assembly**. The quadrant sight assembly, which attaches to the left side of the rifle's carrying handle, enables the grenadier to adjust for elevation and windage (Figure 2-2). This assembly consists of the sight, mounting screw, sight latch, rear sight aperture, sight aperture arm, front sight post, and sight post arm (Figure 2-3).

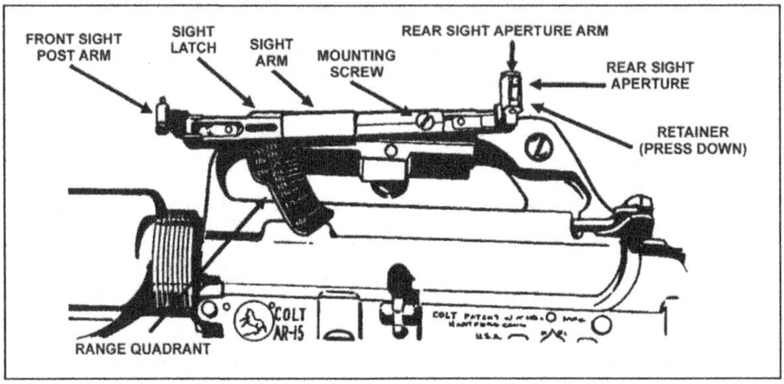

Figure 2-3. Quadrant sight assembly.

(1) ***Clamp, Bracket Assembly, and Mounting Screw***. The clamp and the bracket assembly hold the quadrant sight on the rifle's carrying handle. The mounting screw inserts through the right side of the clamp and into the bracket assembly.

(2) ***Sight Arm and Range Quadrant***. The sight arm mounts both the sight aperture arm (which holds the rear sight aperture) and the sight post arm (which holds the front sight post). This procedure allows the sight to pivot on the range quadrant to the desired range setting. The range quadrant is graduated in 25-meter increments from 50 to 400 m. Applying rearward pressure on the sight latch releases the quadrant sight arm so it can move along the range quadrant. Centering the number in the rear sight aperture selects the desired range. Releasing the sight latch locks the sight in position.

(3) ***Front Sight Post***. The front sight post mounts on the sight post arm by means of a pivot bracket. To prevent damage to the sights, keep the bracket closed when the sights are not in use. Use the sight post as follows to make minor adjustments in elevation when zeroing the launcher:

(a) To decrease elevation, turn the elevation adjustment screw on the sight post clockwise; to increase elevation, turn it counterclockwise.

(b) To move the impact of the projectile 5 meters at a range of 200 m, turn the elevation adjustment screw one full turn--360 degrees. To move the impact of the projectile 2.5 meters at a range of 200 m, turn the elevation adjustment screw one half turn--180 degrees.

(4) ***Rear Sight Aperture***. The rear sight aperture is on the sight aperture arm, which is attached to the rear portion of the quadrant sight arm. Use the rear sight aperture as follows to make minor adjustments in deflection (windage) when zeroing the launcher:

(a) To move the impact to the left, press the rear sight aperture retainer down and move the rear sight aperture away from the barrel; to move to the right, move it toward the barrel.

(b) To move the impact of the projectile 1.5 meter at a range of 200 m, move the rear sight aperture one notch.

c. **Receiver Assembly and Serial Number**. The receiver assembly houses the firing mechanism and ejection system and supports the barrel assembly. On the left side of the receiver assembly is the launcher's serial number (Figure 2-2).

d. **Barrel Assembly**. The barrel assembly holds the cartridges ready for firing and directs the projectile (Figure 2-2).

e. **Barrel Latch**. On the left side of the barrel is a latch that locks the barrel and receiver together (Figure 2-2). To open the barrel, depress the barrel latch and slide the barrel forward.

f. **Leaf Sight Assembly**. The leaf sight assembly is attached to the top of the handguard (Figure 2-4). The leaf sight assembly consists of the sight, its base and mount, an elevation adjustment screw, and a windage adjustment screw. Elevation and windage scales are marked on the mount. The folding, adjustable, open ladder design of the sight permits rapid firing without sight manipulation. The front sight post of the M16-series rifle serves as the front aiming post for the M203 leaf sight.

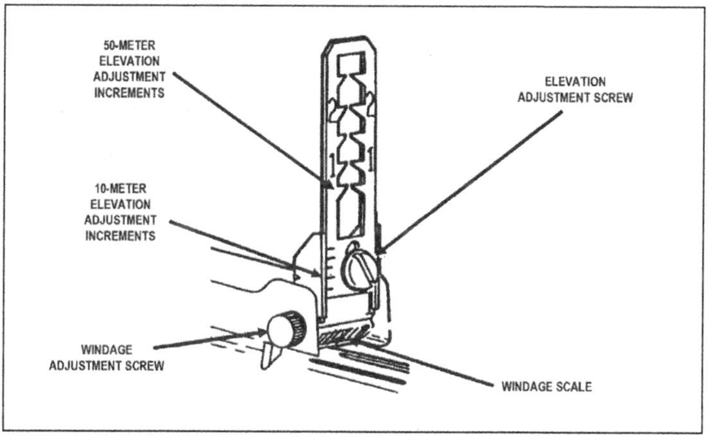

Figure 2-4. Leaf sight assembly.

(1) ***Sight Base***. Two mounting screws permanently attach the sight base to the rifle handguard. When the sight is down, the base protects it from damage.

(2) ***Sight Mount and Sight***. The grenadier uses the sight mount, which is attached to the sight base, to raise or lower the sight. Though the range is not marked on the sight in m, the sight is graduated in 50-meter increments from 50 to 250 m, which are marked with a "1" at 100 meters and a "2" at 200 m.

(3) ***Elevation Adjustment Screw and Elevation Scale***. The screw attaches the sight to its mount. When the screw is loosened, the sight can be moved up or down to make minor adjustments in elevation during the zeroing procedure. The rim of a 40-mm cartridge case is useful for turning the screw. Raising the sight increases the range;

lowering the sight decreases the range. The elevation scale consists of five lines spaced equally on the sight. The index line is to the left of the sight. Moving the sight one increment moves the impact of the projectile 10 meters in elevation at a range of 200 meters.

(4) ***Windage Screw and Windage Scale***. The firer uses the knob on the left end of the windage screw to make minor deflection adjustments during the zeroing. The scale has a zero line in the center. Two vertical lines are located on either side of and equidistant from the zero line. At a range of 200 meters, turning the knob on the windage scale one increment to the *left* moves the impact of the projectile 1.5 meter in the opposite direction, that is, to the *right*.

DANGER

NEVER **USE HIGH-EXPLOSIVE AMMO TO ZERO AT 50 METERS OR LESS--AT THAT RANGE, FRAGMENTATION COULD KILL YOU.**

TO REMIND YOU OF THIS, THE 50-METER MARK ON THE LEAF SIGHT BLADE IS PAINTED RED.

f. **Trigger Guard**. The trigger guard protects the trigger (A, Figure 2-5). Depressing the rear portion of the trigger guard rotates it down and away from the magazine well of the rifle. This gives the firer room to wear gloves or mittens while firing, if needed (B, Figure 2-5).

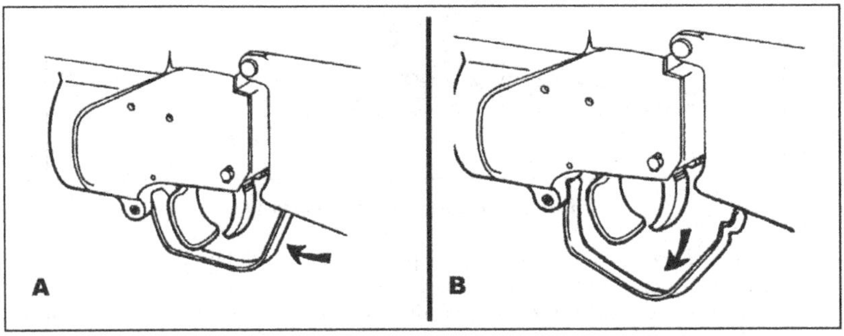

Figure 2-5. Trigger guard.

g. **Safety**. The safety is inside the trigger guard, just in front of the trigger. For the launcher to fire, the safety must be forward. When the safety is rearward, the launcher is on "S." The grenadier adjusts the safety manually (Figure 2-6).

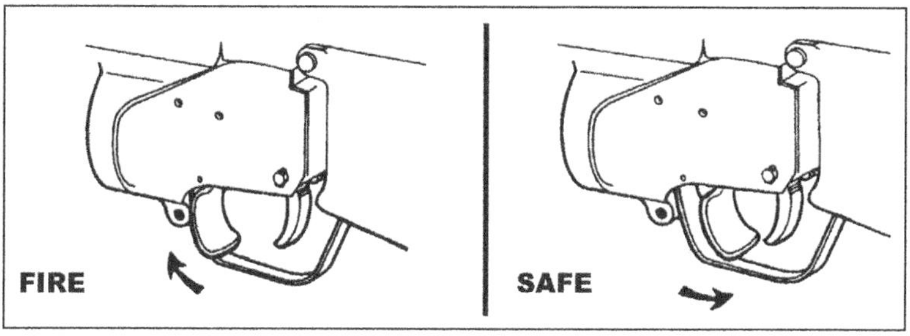

Figure 2-6. Safety.

2-4. AMMUNITION

The M203 grenade launcher uses several fixed-, low-velocity, 40-mm rounds. These include high-explosive, illuminating, signaling, CS, training, and multipurpose ammunition. This paragraph covers the most commonly used rounds.

WARNING

If fired into snow or mud, 40-mm rounds might hit too softly to detonate. An undetonated round might then explode when stepped on or driven over.

When training in snow or mud, avoid this hazard by firing only TP rounds.

a. **Types, Characteristics, and Capabilities**. Figure 2-7 shows and Table 2-1 compares the M203 rounds (TM 43-0001-28 provides more details.)

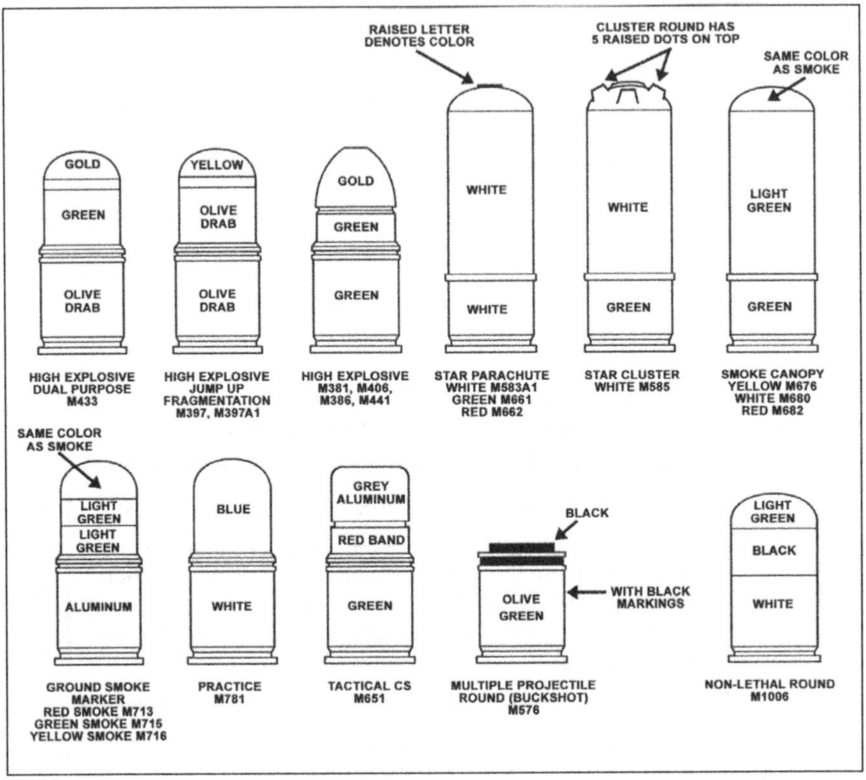

Figure 2-7. Cartridges for the M203 grenade launcher.

MODEL	DESCRIPTION	APPLICATION/EFFECTS
High Explosive, Dual-Purpose Round M433: DODAC 1310-B546 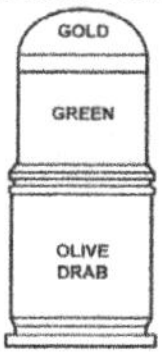 	Ogive: Gold-colored steel cup Body: Green Skirt: Olive drab aluminum Markings: White Length: 10.29 cm (4.05 in) Weight: 0.23 kg (0.51 lb)	This round has a small slug (1/4 inch diameter) that penetrates at least 5 cm (2 in) of steel armor when fired straight (flat, that is, at 0 degree obliquity) within 150 m. When fired at a point target, it arms between 14 and 27 m (46 to 89 ft). It can cause-- • Casualties within 165 m. • Incapacitation within 15 m. • Death within 5 m.

WARNING

In training, seek cover before firing an HE round at a target within 165 meters.

Also in training, never engage targets within a 165 meter radius, that is, within 165 meters in any direction from unprotected friendly troops.

MODEL	DESCRIPTION	APPLICATION/EFFECTS
High-Explosive Jump-Up Fragmentation M397 and M397A1 DODAC 1310-B569 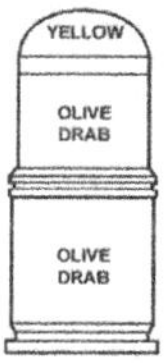 	Ogive: Yellow steel ball Skirt: Olive drab aluminum Filler: Explosive Markings: Yellow Length: 4.05 in. Weight: 0.51 lb	These rounds are designed to inflict personnel casualties using an air burst effect. It arms between 14 and 27 meters (46 and 89 feet). It detonates and fragments about 5 feet above the ground impact point.
High-Explosive M381: DODAC 1310-B568 M441: DODAC 1310-B575 M386: DODAC 1310-B574 M406: DODAC 1310-B568 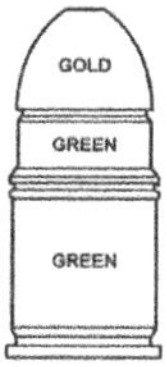 	Ogive: Gold Body: Green Skirt: Olive drab aluminum with attached steel projectile Markings: Gold Length: 3.89 in. Weight: M381: 5.03 lb M441: 5.03 lb M386: 0.50 lb M406: 0.50 lb	These rounds produce a ground burst that causes casualties within a 165 meter radius. The kill radius is 5 meters. The M381 and M441 rounds arm within 2.4 to 3 meters (8 to 10 feet). The M386 and M406 rounds arm between 14 and 27 meters (46 to 89 feet).

Table 2-1. Comparison of M203 rounds.

MODEL	DESCRIPTION	APPLICATION/EFFECTS
Star Parachute Rounds M583A1 (WHITE) DODAC 1310-B531 90,000 CANDLEPOWER M661 (GREEN) DODAC 1310-B504 8,000 CANDLEPOWER M662 (RED) DODAC 1310-B505 20,000 CANDLEPOWER 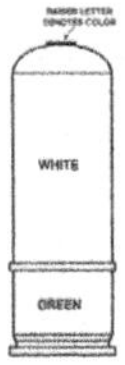	Cap: White, with raised letters indicating parachute color Body: White impact or bar alloy aluminum Skirt: Green Markings: Black Length: 13.39 cm (5.27 in) Weight: 0.22 kg (0.49 lb)	These rounds are used for illumination and signals, and are lighter and more accurate than comparable handheld signal rounds. The parachute attached to the round deploys upon ejection to lower the candle at 7 feet per second. The candle burns for about 40 seconds.
White Star Cluster M585 DODAC 1310-B536 	Cap: Plastic, with five raised dots for night identification. Body: White impact or bar aluminum alloy Skirt: Green Markings: Black Length: 13.38 cm (5.27 in) Weight: 0.19 kg (0.41 lb)	This round is used for illumination and signals. It is lighter and more accurate than comparable handheld signal rounds. The individual stars burn for about 7 seconds during free fall.
Ground Marker (Smoke) Rounds M713, (RED) DODAC 1310-B506 M715 (GREEN) DODAC 1310-B508 M716 (YELLOW) DODAC 1310-B509 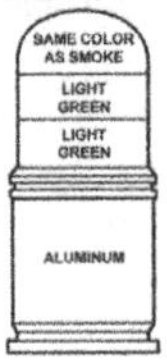	Ogive: Same as smoke color Body: Light green impact aluminum Skirt: Aluminum Markings: Black Length: 3.91 in. Weight: 0.49 lb	This round is used to mark soldiers' locations for others to see from the ground or the air. These rounds arm between 15 and 45 meters. If a fuze fails to function on impact, the output mixture in the front end of the delay casing backs up the impact feature.

Table 2-1. Comparison of M203 rounds (continued).

MODEL	DESCRIPTION	APPLICATION/EFFECTS
Smoke Canopy Rounds M676 (Yellow) DODAC 1310-B475 M680 (White) DODAC 1310-B477 M682 (RED) DODAC 1310-B479  	Ogive: Same as color of smoke Body: Light green impact aluminum Skirt: Aluminum Markings: Black Length: 5.212 in. Weight: 0.48 lb	This round is used to mark soldiers' locations for others to see from the ground or the air. These rounds arm between 15 and 45 meters. Burst height is 91 meters.
Practice Round M781 DODAC 1310-B519 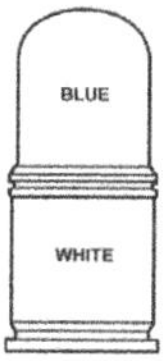 	Windshield: Blue plastic Body: Zinc or aluminum Skirt: White Markings: White Length: 10.29 cm (4.05 in) Weight: 0.22 kg (0.48 lb)	Round produces a yellow or orange signature on impact.
Tactical CS M651 DODAC 1310-B567 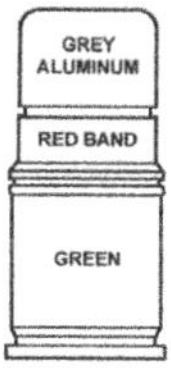 	Ogive: Grey aluminum Band: Red Skirt: Green Markings: Black markings Length: 11.43 cm (4.50 in) Weight: 0.22 kg (0.48 lb)	This multipurpose round works best in riot control and urban operations. It arms between 10 to 30 meters (33 to 99 ft) and produces a white cloud of CS gas on impact.
Multiple Projectile (Buckshot) M576: DODAC 1310-B534  	Ogive: NA Body: Olive drab Skirt: NA Markings: Black Length: 6.72 cm (2.65 in) Weight: 0.12 kg (0.25 lb)	This multipurpose round is best in thick, vegetated areas or for use in room clearing. It contains at least 20 pellets, which cast a cone of fire 30 meters wide and 30 meters high, and which travel at 269 meters per second. Buckshot rounds should be aimed at the foot of the target. This round has no mechanical type fuze.

Table 2-1. Comparison of M203 rounds (continued).

MODEL	DESCRIPTION	APPLICATION/EFFECTS
Nonlethal Round M1006: DODAC 1310-BA06 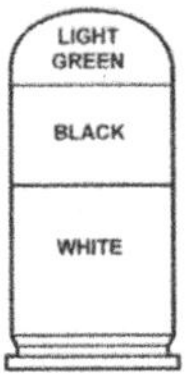	Nose: Green foam rubber Body: Black, high-density plastic Skirt: NA Other: Projectile assembly Cartridge case assembly Length: 3.95 in. Weight: 0.15 lb	This round is used to incapacitate personnel without penetrating the body. It is used to control riots, police, and keep the peace.

Table 2-1. Comparison of M203 rounds (continued).

b. **Storage**. Ammunition should be stored under cover. If this is not possible, it should be stored at least 15 centimeters (6 inches) aboveground, and covered with a double layer of tarpaulins. Tarpaulins should protect the ammunition, but allow ventilation. Trenches should be dug to prevent water from flowing under the ammunition.

c. **Care, Handling, and Preservation**. Ammunition containers should remain tightly closed until the ammunition is needed. Ammunition removed from the airtight container is likely to corrode, particularly in damp climates. Soldiers must also take the following precautions:

(1) ***Protect Ammunition from Mud, Dirt, and Water***. If the ammunition gets wet or dirty, wipe it off before use. Also, wipe off lightly corroded cartridges as soon as you find the corrosion. Never fire heavily corroded or dented projectiles or those with loose parts or particles.

WARNING

Never fire heavily corroded or dented projectiles, or those with loose parts or particles.

(2) ***Avoid Exposing Ammunition to the Direct Rays of the Sun.*** Hot powder can cause excessive pressure when the round is fired.

(3) ***Do Not Lubricate Ammunition***. This can cause dust and other abrasives to collect on it and damage the operating parts of the launcher.

d. **Packaging**. Ammunition packaging varies according to the type of ammunition:

(1) ***HE, HEDP, and TP***. Each wire-bound box of HE and HEDP ammunition contains 1 fiber box with 12 bandoleers of 6 rounds each, for a total of 72 rounds. Each wire-bound box of M781 practice rounds contains 4 fiber boxes of 25 rounds each, for a total of 100 rounds.

(2) ***Smoke and Cluster Ammunition***. Each wire-bound box of smoke and cluster ammunition contains 2 cans with 22 rounds each, for a total of 44 rounds.

(3) ***CS Ammunition***. Each box of CS ammunition contains 2 cans with 4 bandoleers of 6 rounds each, for a total of 48 rounds.

(4) ***Buckshot***. Each box of buckshot ammunition contains 12 bandoleers of 6 rounds each, for a total of 72 rounds.

(5) ***Nonlethal Ammunition.*** Each wire-bound wooden box contains 22 rounds per M2A1 metal box, for a total of 44 rounds.

3-5. CLEARING PROCEDURES

The soldier must clear the weapon before performing maintenance on it. FM 3-22.9 (23-9) provides instructions for clearing an M16-series rifle. To clear the grenade launcher--

a. Push in the release button and pull the barrel forward.

b. Watch to see if a round extracts.

c. Place the safety on "S."

d. Inspect the breech to ensure a round is not present.

e. Pull the barrel to the rear until it clicks. This cocks the weapon.

2--6. GENERAL DISASSEMBLY

As he disassembles the weapon, the soldier places each part on a clean, flat surface such as a table, shelter half, or disassembly mat. This aids in reassembly and simplifies the task of keeping up with the parts. The soldier will later reassemble the grenade launcher in the reverse order that he disassembled it. (Only ordnance personnel may disassemble the grenade launcher beyond the steps described here.) To disassemble the weapon--

a. Loosen the mounting screw and remove the quadrant sight assembly from the carrying handle of the M16-series rifle (Figure 2-8).

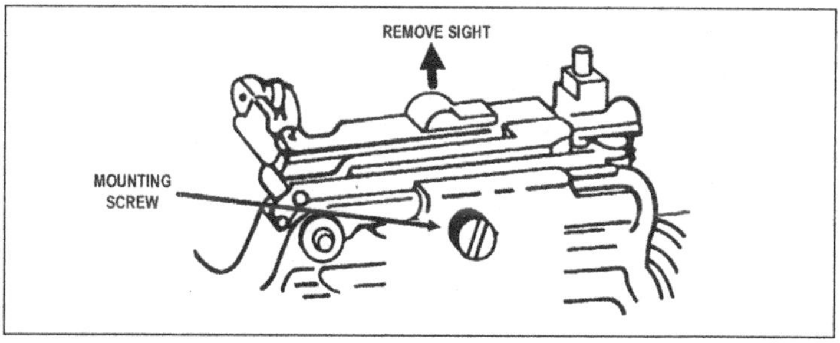

Figure 2-8. Removal of the quadrant sight assembly.

b. Remove the barrel assembly and handguard assembly, in either order:

(1) ***Barrel Assembly First.*** Push the barrel latch and move the barrel forward until it hits the barrel stop. On the left side of the handguard, insert a cleaning rod into the fourth hole back from the muzzle, depress the barrel stop, and slide the barrel forward and off (Figure 2-9).

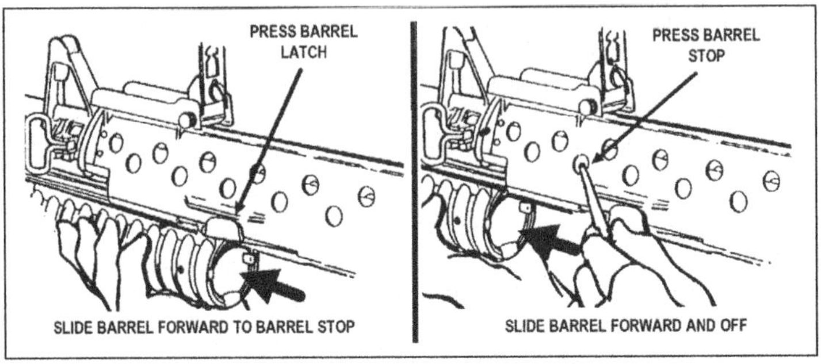

Figure 2- 9. Removal of the barrel assembly before the handguard assembly.

(2) ***Handguard Assembly First***. Pull back on the M16's slip ring and remove the handguard by pulling it up and back. Push the barrel latch and move the barrel forward until it hits the barrel stop. Use a cleaning rod to depress the barrel stop and slide the barrel forward and off (Figure 2-10).

> **CAUTION**
>
> Never use a screwdriver or any other tool when removing the hand guard assembly. Doing so could damage the hand guard assembly or slip ring.

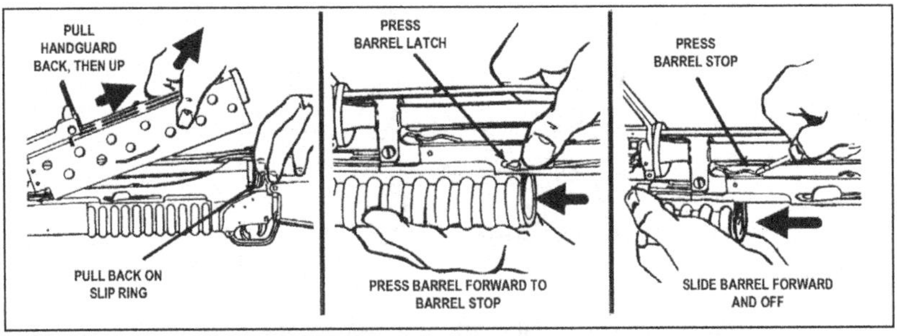

Figure 2-10. Removal of the handguard assembly before the barrel assembly.

2-7. CLEANING AND LUBRICATING PROCEDURES

After firing the grenade launcher, or if it has been idle for a long time, the soldier must clean and lubricate it as follows:

a. **Bore**. Attach a clean, dry rag to the thong, and then thoroughly moisten the rag with CLP. Pull the rag through the bore several times. Attach the bore brush to the thong, pull it through the bore several times, and follow this with more rags moistened with CLP (Figure 2-11). Pull dry rags through the bore, and inspect each rag as it is removed. The bore is clean when a dry rag comes out clean. Finally, pull a rag lightly moistened with CLP through the bore to leave a light coat of lubricant inside the barrel.

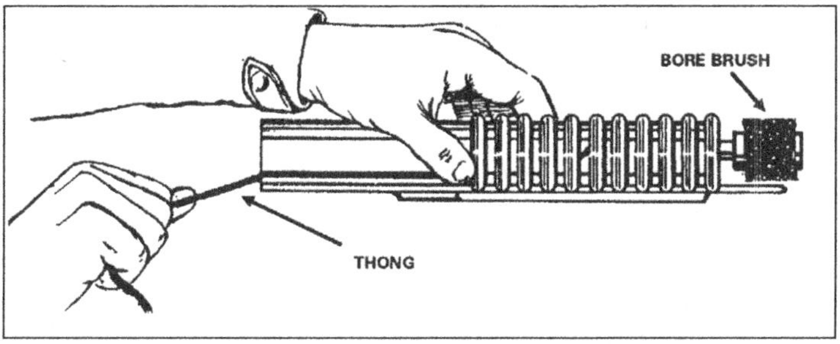

Figure 2-11. Cleaning of the barrel with thong and bore brush.

b. **Breech Insert**. Clean the face of the breech insert with a patch and CLP. Remove the CLP with dry rags; then lubricate the breech with a new, light coat of CLP.

c. **Other Parts**. Use a brush and dry rags to clean all the other parts and surfaces. After cleaning, apply a light coat of CLP to the outside of the launcher.

d. **Safety Mechanism**. Clean the safety mechanism properly with CLP; then lubricate it with CLP.

e. **Special Lubrication Requirements**. Lubricate the grenade launcher only with CLP and IAW the following environmental guidelines:

(1) ***Extreme Heat***. Lubricate with CLP, grade 2.

(2) ***Damp or Salty Air***. Clean the weapon and apply CLP, grade 2, frequently.

(3) ***Sandy or Dusty Air***. Clean the weapon and apply CLP, grade 2, frequently. Remove excess CLP with a rag after each application.

(4) ***Temperatures Below Freezing***. When the weapon is brought in from a cold area to a warm area, keep it wrapped in a parka or blanket, and allow it to reach room temperature gradually. If condensation forms on the weapon, dry and lubricate it at room temperature with CLP, grade 2, before returning it to cold weather. Otherwise, ice will form inside the mechanism.

NOTE: Although CLP provides sufficient lubrication down to -35°F (-37°C), it will only flow from its 1/2-ounce bottle, at 0°F (-17°C) and above.

2-8. INSPECTION

Inspection begins with the weapon already disassembled into its major groups or assemblies. Parts with shiny surfaces are serviceable. The following parts of the weapon and related equipment are inspected IAW TM 9-1010-221-10:

a. **All Parts**. Check for wear and damage, including burrs, scratches, and nicks.

b. **Handguard**. Check for cracks, dents, or distortion that prevents its firm attachment to the rifle.

c. **Leaf Sight Assembly**. Check for bent or damaged parts, rust or corrosion, and legible markings.

d. **Barrel**. Check for cracks or dents.

e. **Cartridge and Retainers**. Check for breakage, bends, chips, or missing parts.

NOTE: Take any unserviceable part to the armorer, who will decide if it is serviceable and replace parts as needed.

2-9. GENERAL ASSEMBLY

The soldier assembles the grenade launcher in the reverse order of disassembly.

a. Install the barrel by pressing the barrel stop and sliding the barrel into the receiver (Figure 2-12).

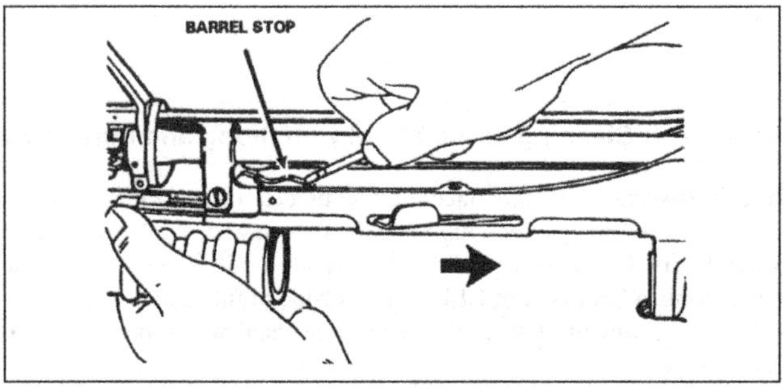

Figure 2-12. Installation of the barrel.

b. Lock the barrel by moving it rearward until it closes with a click (Figure 2-13).

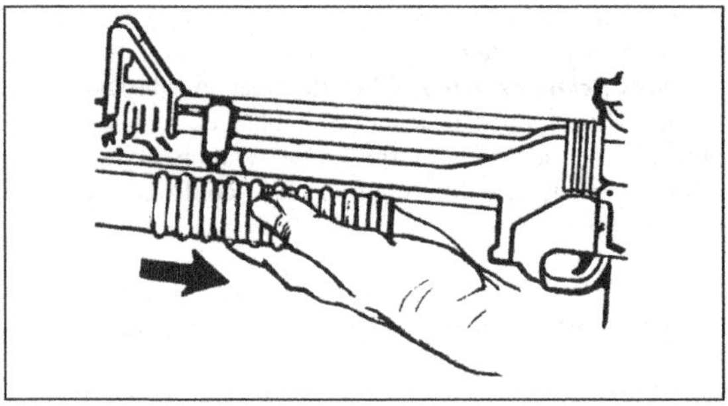

Figure 2-13. Locking of the barrel.

c. Install the handguard and secure it with the slip ring (Figure 2-14).

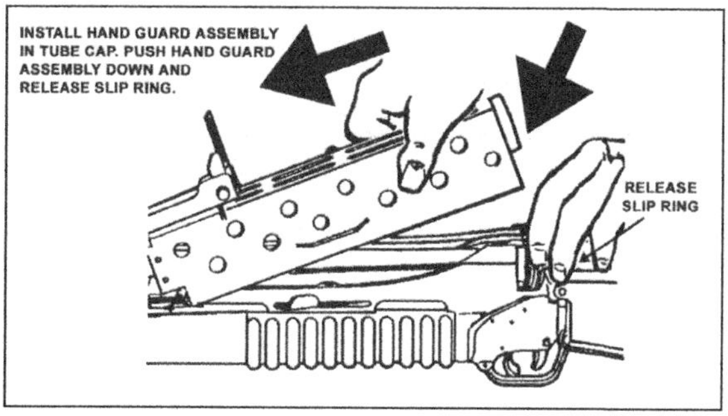

Figure 2-14. Installation and securing of the handguard.

d. Install the quadrant sight assembly (Figure 2-15).

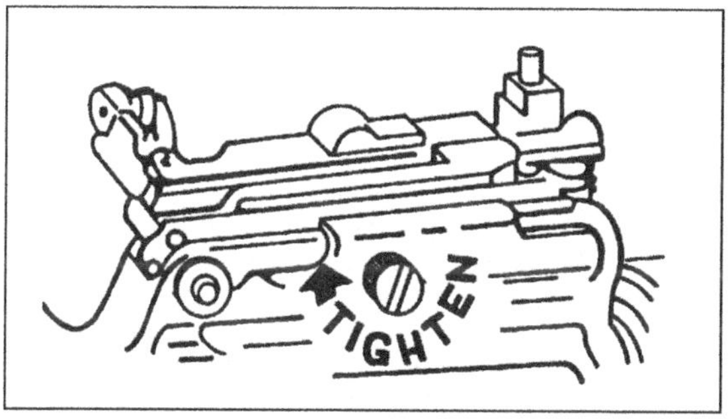

Figure 2-15. Installation of the quadrant sight assembly.

e. Perform a function check to ensure that the grenade launcher has been assembled correctly. Notify the unit armorer at once if the launcher fails to function. Conduct the function check in this order:

(1) Check the proper operation of the sear. Cock the launcher, pull the trigger, and continue to hold the trigger. When you pull the trigger, you release the firing pin and should hear a metallic click The firing pin should release with a metallic click. Hold the trigger to the rear and cock the launcher again. Release the trigger, then pull. The firing pin should again release.

WARNING

If the sear malfunctions, the launcher could fire without your pulling the trigger.

(2) Check the safety by pulling the trigger in both the "S" and "F" positions. You must cock the launcher before you can place the safety in the "S" position.

(3) Check the windage adjustment screw on the leaf sight assembly for proper operation.

(4) Move the elevation adjustment screw only if the weapon has been zeroed.

(5) Move the barrel forward and back to be sure the barrel stop and barrel latch function.

2-10. CARE AND HANDLING PROCEDURES

Certain steps must be taken before, during, and after firing to properly maintain the grenade launcher.

a. **Before firing**.

- Wipe the bore dry.
- Inspect the weapon as outlined in the operator's technical manual.
- Ensure the weapon is properly lubricated.

b. **During firing**.

- Periodically inspect the weapon to ensure that it is lubricated.
- When malfunctions or stoppages occur, see the procedures in Chapter 4.

c. **During CBRN Conditions**. If contamination is anticipated, the soldier should apply CLP to all outer metal surfaces of the weapon. Ammunition, however, should never be lubricated. The soldier should keep the weapon covered as much as possible. If the weapon is contaminated, he should first decontaminate it IAW FM 3-3 and FM 3-11.4, and then he should clean and lubricate it.

2-11. DECONTAMINATION

Leaders must try to reduce the penetration of contaminants and exposure to them. Contaminated material is disposed of IAW SOP.

a. **Chemical**. Use towelettes from the M258A1 kit to wipe off the weapon. If these are unavailable, wash the weapon with hot soapy water.

b. **Biological**. Use towelettes from the M258A1 kit to wipe off the weapon. If these are unavailable, wash with soap and water.

c. **Radiological or Nuclear.** Wipe off the weapon with warm soapy water. Otherwise, use towelettes or rags (FM 3-11.5).

*CHAPTER 3

OPERATION AND FUNCTION

This chapter discusses the operation and function of the M203 grenade launcher.

3-1. OPERATION

The grenadier's operations include loading, unloading, and firing the weapon. The weapon uses a high-low propulsion system to fire a 40-mm round. The firing pin strikes the primer, whose flash ignites the propellant in the brass powder-charge cup inside the high-pressure chamber. The burning propellant produces 12,000 psi chamber pressure, which ruptures the brass powder-charge cup at the vent holes and allows the gases to escape to the low-pressure chamber in the cartridge case. There the pressure drops to 3,000 psi and propels the grenade from the muzzle at a velocity of 250 fps. The grenade's 3,700-rpm right-hand spin stabilizes it during flight and applies enough rotational force to arm the fuze. The weapon is unloaded with the barrel open and fired from a closed bolt. It must be cocked before it can be placed on safe.

3-2. LOADING OF M203 WITH THE SAFETY ON ("S")

When the safety is on "S," the grenadier loads the M203 as follows (Figure 3-1):

a. Keep the safety on "S" until ready to fire.

b. Depress the barrel latch.

c. Slide the barrel assembly forward. This automatically extracts a round and cocks the weapon.

CAUTION

Before loading the M203, make sure that the bore and chamber are clean and dry.

d. Insert clean, dry, undented ammunition into the chamber.

e. Slide the barrel rearward until it locks with an audible click.

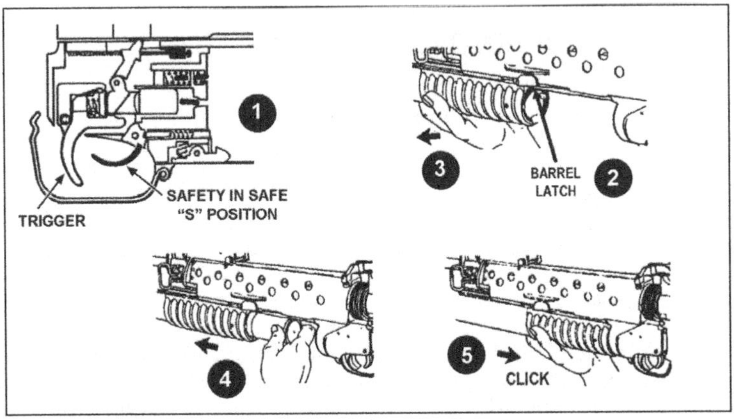

Figure 3-1. Loading the M203 with the safety on ("S").

3-3. LOADING OF M203 WITH THE SAFETY OFF ("F")

When the safety is off, the grenadier loads the M203 as follows (Figure 3-2):

a. Depress the barrel latch.

b. Slide the barrel assembly forward. This automatically extracts a round and cocks the weapon.

c. Place the weapon on safe ("S").

> **CAUTION**
> Before loading ensure bore and chamber is clean and dry.

d. Insert clean, dry, undented ammunition into the chamber.

e. Slide the barrel rearward until it locks with an audible click.

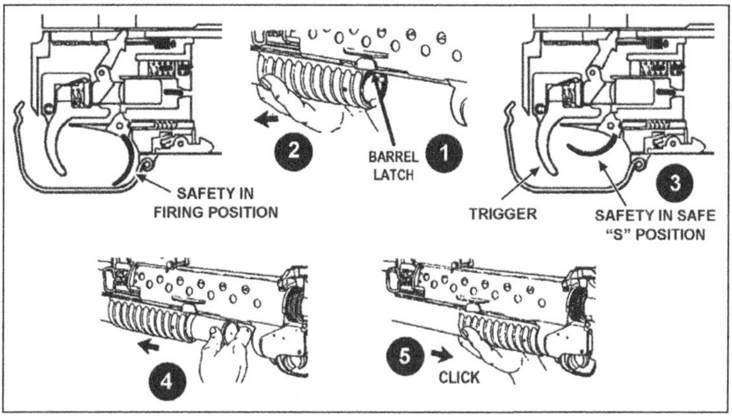

Figure 3-2. Loading the M203 with the safety off ("F").

> **WARNING**
>
> **Keep the muzzle pointed downrange and clear of all Soldiers. Use the correct ammunition:**
>
> - **Never fire pyrotechnic ammunition made for the AN-M8 pyrotechnic pistol in the M203.**
> - **Never load aircraft ammunition M384 (HE) or M385 practice.**
> - **Never use high-velocity 40-mm ammunition designated for other 40-mm weapons such as the MK 19. High-velocity rounds are longer than those used in the M203 and might cause this weapon to explode, resulting in serious injury.**

3-4. UNLOADING OF M203

To unload the grenade launcher, the grenadier must first depress the barrel latch and move the barrel forward. The cartridge case or round should automatically eject. If the case sticks, the grenadier taps it with a cleaning rod to remove it (Figure 3-3). He places the weapon on safe ("S"), then slides the barrel rearward, locking it to the breech.

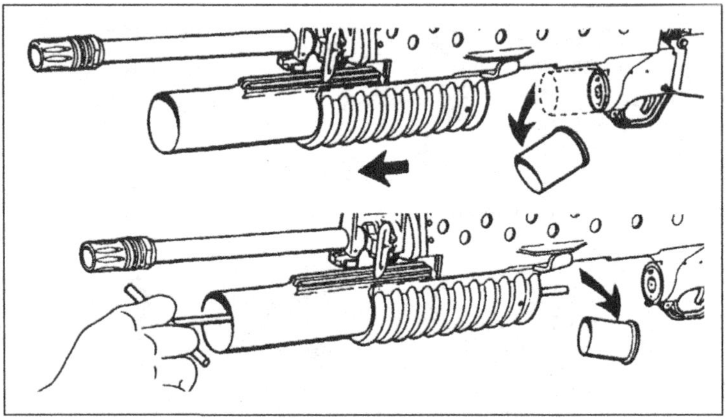

Figure 3-3. Unloading the M203 grenade launcher.

WARNING

To unload an unfired weapon, avoid detonation by either--

- **Catching the ejected round or**
- **Holding the weapon close to the ground to reduce the distance the round can fall.**

3-5. CYCLE OF FUNCTIONING

Knowing the M203's cycle of functioning from loading to firing helps grenadiers recognize and correct stoppages. Many of the actions described in this chapter occur at once, but here they are explained separately.

a. **Unlocking**. The cycle begins when the grenadier depresses the barrel latch to unlock the barrel assembly and slides the barrel assembly forward (Figure 3-4).

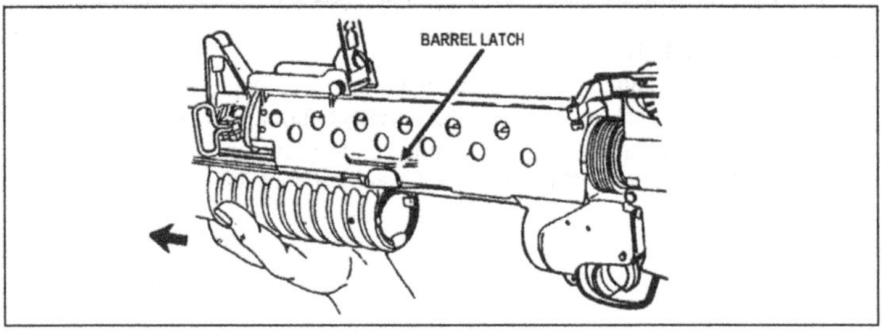

Figure 3-4. Unlocking the barrel assembly.

b. **Cocking**. The grenadier moves the barrel assembly forward, then backward, to cock the weapon. As the barrel assembly moves, it takes with it the barrel extension. Their movement causes the following to occur:

(1) The cocking lever is forced down as the barrel assembly and barrel extension, which are interlocked with the cocking lever, move forward.

(2) The movement of the cocking lever forces the spring-loaded firing pin to the rear.

(3) The spring-loaded follower also moves forward with the barrel extension.

(4) The barrel assembly continues forward, disengaging the barrel extension from the cocking lever. The cocking lever is then held down by the follower.

(5) When the grenadier begins to move the barrel assembly back to the rear, this forces the follower to the rear.

(6) The cocking lever again engages the barrel extension, which causes the firing pin to move slightly forward and engage the primary trigger sear. This cocks the weapon (Figure 3-5).

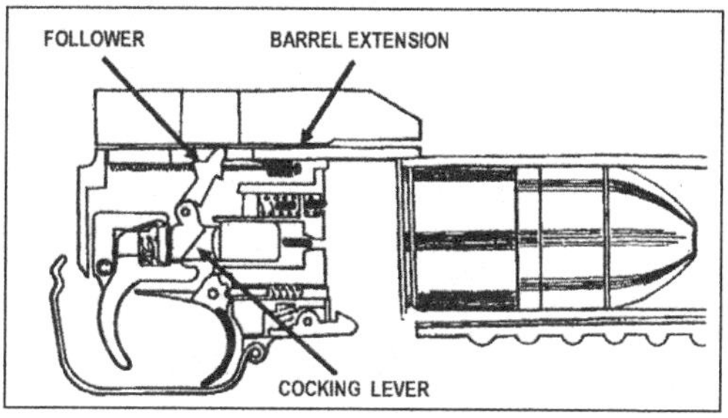

Figure 3-5 . Cocking the M203 grenade launcher.

c. **Extracting**. Extracting and cocking occur at the same time. As the grenadier opens the barrel assembly, a spring-loaded extractor keeps the live round or spent cartridge case seated against the receiver until the barrel clears the cartridge case (Figure 3-6).

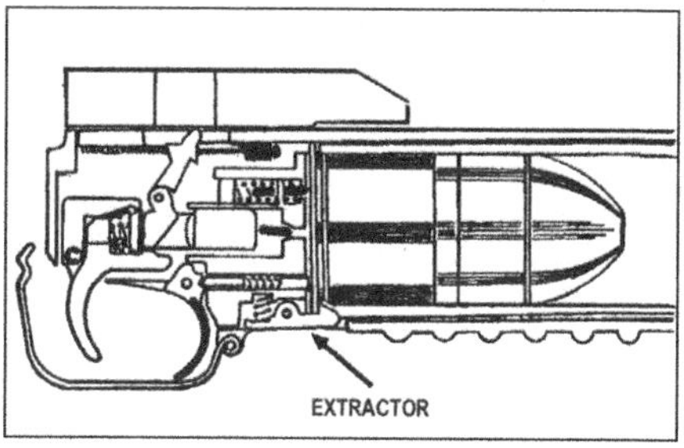

Figure 3-6 . Extracting the round or cartridge case.

d. **Ejecting**. The spring-loaded ejector pushes the live round or spent cartridge case from the barrel assembly (Figure 3-7).

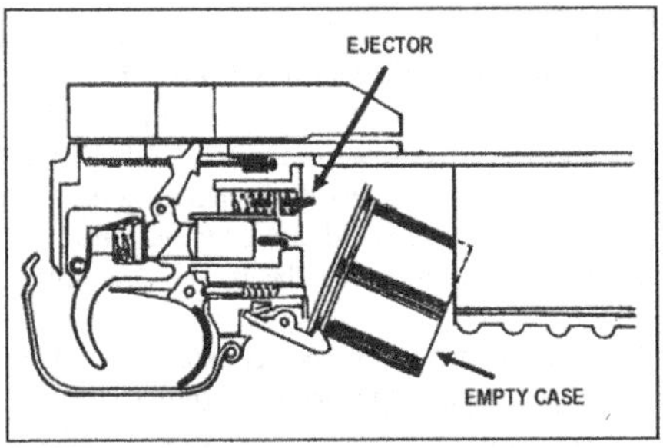

Figure 3-7 . Ejecting the round or cartridge case.

e. **Loading**. With the barrel assembly open, the grenadier inserts a round into the breech end of the barrel (Figure 3-8).

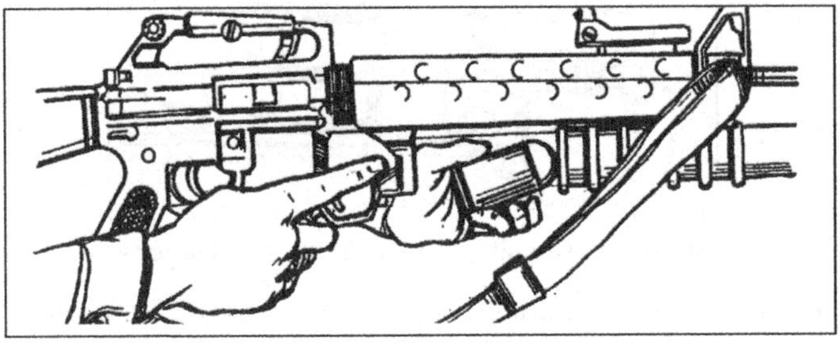

Figure 3-8 . Loading the M203 grenade launcher.

f. **Chambering**. As the grenadier closes the breech end of the barrel assembly, the extractor contacts the rim of the cartridge and seats (chambers) the round firmly (Figure 3-9).

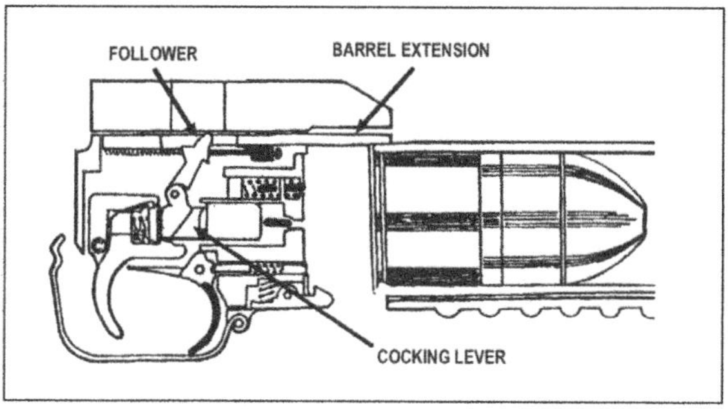

Figure 3-9 . Chambering a round.

g. **Locking**. As the barrel assembly closes, the barrel latch engages it. The cocking lever engages the barrel extension so that it cannot move forward along the receiver assembly.

h. **Firing**. When the grenadier pulls the trigger, the primary trigger sear disengages from the bottom sear surface of the firing pin. This releases the spring-driven firing pin, forcing it forward against the cartridge primer (Figure 3-10).

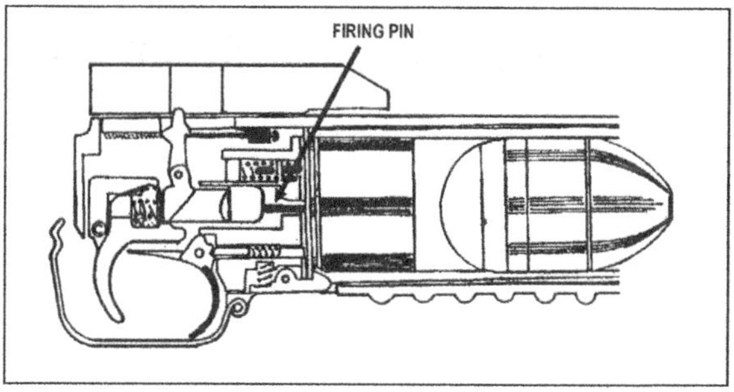

Figure 3-10. Firing the M203 grenade launcher.

This page intentionally left blank.

CHAPTER 4
PERFORMANCE PROBLEMS AND DESTRUCTION

This chapter identifies some of the problems that can cause the M203 grenade launcher to perform incorrectly. It also explains how to identify unserviceable parts and how to destroy the weapon when authorized to do so.

4-1. MALFUNCTIONS

A malfunction occurs when a mechanical failure prevents the weapon from firing properly. Neither defective ammunition nor improper operation of the weapon by the firer is a malfunction. The weapon should be cleaned, lubricated, and retried. If it still fails to function, it should be turned in to the unit armorer. Table 4-1 shows probable causes and corrective action for each type of malfunction.

Malfunction	Probable Cause	Corrective Action
Failure to cock	Broken sear	Notify unit maintenance
	Improper assembly of cocking lever	
	Loose, broken, or missing cocking lever spring pin	
Failure to lock	Excess plastic on breech end of barrel assembly	

Table 4-1. Malfunctions.

4-2. STOPPAGES

A stoppage is an unintentional interruption in the cycle of operation or functioning that may be cleared by immediate action. A stoppage is classified by its relationship to the cycle of functioning. Table 4-2 on page 4-2 shows the types of stoppages.

<table>
<tr><th>Stoppage</th><th>Probable Cause</th><th>Corrective Action</th></tr>
<tr><td rowspan="9">Failure to fire</td><td>Safety on</td><td>Place in fire position</td></tr>
<tr><td>Empty chamber</td><td>Load</td></tr>
<tr><td>Faulty ammunition</td><td>Reload</td></tr>
<tr><td>Water or excess lubricant in firing pin well</td><td>Hand cycle weapon several times, to include pulling the trigger</td></tr>
<tr><td>Worn or broken firing pin</td><td>Notify unit maintenance</td></tr>
<tr><td>Dirt or residue in firing pin recess</td><td>Clean</td></tr>
<tr><td>Blurred sear or firing pin</td><td rowspan="4">Notify unit maintenance</td></tr>
<tr><td>Dirty firing pin well opening</td></tr>
<tr><td>Weak or broken firing pin spring</td></tr>
<tr><td rowspan="2">Failure to extract</td><td>Defective extractor on spring or spring pin</td></tr>
<tr><td>Ruptured cartridge case</td><td>Remove from barrel</td></tr>
<tr><td>Failure to eject</td><td>Worn, broken, or missing ejector spring or retainer</td><td>Notify unit maintenance</td></tr>
<tr><td rowspan="2">Failure to chamber</td><td>Faulty ammunition</td><td>Reload</td></tr>
<tr><td>Dirty chamber</td><td>Clean bore and chamber</td></tr>
<tr><td>Safety fails to stay in position</td><td>Missing spring pin or broken or worn safety</td><td>Notify unit maintenance</td></tr>
</table>

Table 4-2. Stoppages.

*4-3. IMMEDIATE ACTION

Immediate action refers to anything a Soldier does to reduce a stoppage without taking time to look for the cause. Immediate action should be taken in the event of either a hangfire or misfire. Either can be caused by an ammunition defect or by a faulty firing mechanism. Any failure to fire must be considered a hangfire until that possibility is eliminated.

a. What is the difference between a hangfire, a misfire, and a stoppage?

(1) A *hangfire* is a delay in the functioning of the round's propelling charge explosive train at the time of firing. The length of this delay is unpredictable, but in most cases, it ranges between a split second and 30 seconds. Such a delay in the functioning of the round could result from the presence of excess oil or grease, grit, sand, frost, or ice.

(2) A *misfire* is a complete failure of the weapon to fire. A misfire in itself is not dangerous, but because it cannot be immediately distinguished from a hangfire, it must be considered to be a hangfire until proven otherwise.

(3) A *stoppage* may be caused by a hangfire, the following precautions must be observed until the round has been removed from the weapon and the cause of the failure determined:

(a) Keep the M203 pointed downrange or at the target and keep everyone clear of its muzzle. If the stoppage occurs during training, shout MISFIRE and clear the area of any Soldiers not needed for the operation.

(b) Wait 30 seconds from the time of the failure before opening the barrel assembly. Once the 30 seconds has elapsed, open the barrel assembly slowly to about one-half inch or until you hear a metallic click. As soon as you hear the click, place the weapon on safe. You are now ready to perform the unloading procedure.

(c) After removing the round from the receiver, determine whether the round or the firing mechanism is defective. Examine the primer to see if it is dented. If the primer is dented, separate the round from other ammunition until it can be disposed of properly. However, if the primer is not dented, the firing mechanism is at fault. Once the cause of the failure to fire has been corrected, the round may be reloaded and fired.

WARNING

If you are unloading a weapon that has not been fired, avoid detonation either by catching the ejected round or by holding the weapon close to the ground to reduce the distance the round can fall.

4-4. REMEDIAL ACTION

Remedial action is any action taken by the gunner to restore his weapon to operational condition. Take remedial action only if immediate action does not remedy the problem.

4-5. DESTRUCTION PROCEDURES

Destruction of any military weapon is authorized only as a last resort to prevent the enemy from capturing or using it. This paragraph discusses planning for destruction, priorities and methods of destruction, and degree of damage. In combat situations, the commander has the authority to destroy weapons, but he must report doing so through channels.

a. **Planning**. SOPs for all units should contain a plan for destroying equipment. Having such a plan ensures that the damage is effective enough to deny use of the equipment to the enemy. The plan must be flexible enough in its designation of time, equipment, and personnel to meet any situation.

b. **Priorities of Destruction**. When lack of time prevents them from completely destroying equipment, Soldiers must destroy the same essential parts on all like equipment. The order in which the parts should be destroyed (priority of destruction) follows:

(1) Bolt assembly (M16) and breech mechanism (M203).

(2) Barrels (both M16 and M203).

(3) Sights or sighting equipment (including nightsight).

(4) Optics mount.

c. **Methods of Destruction**. Equipment may be destroyed by any of several methods. The commander must use his imagination and resourcefulness to select the best method of destruction based on the facilities available. Time is usually critical. The methods of destruction are as follows:

(1) ***Mechanical***. Use an axe, pick, sledgehammer, crowbar, or other heavy implement.

(2) ***Burning***. Use gasoline, oil, incendiary grenades, other flammables, or a welding or cutting torch.

(3) ***Demolition***. Use suitable explosives or ammunition or, as a last resort, hand grenades.

(4) ***Disposal***. Bury essential parts, dump them in streams, or scatter them so widely that recovering them would be impossible.

d. **Degree of Damage**. The method of destruction used must damage equipment and essential spare parts to the extent that they cannot be restored to usable condition in the combat zone, either by repair or by cannibalization.

CHAPTER 5

MARKSMANSHIP TRAINING

**Marksmanship training is conducted in three phases. Every phase has the same three objectives: to teach each grenadier--*

- *To hit the target accurately with the first round,*
- *To adjust fire, and*
- *To do both quickly.*

*This chapter discusses the first two phases: **preliminary marksmanship training**, which develops nonfiring individual skill proficiency (Section I), and **basic gunnery** (Section II). In these two phases, the Soldier learns to apply the fundamentals of gunnery and to zero the M203 during qualification exercises in day, CBRN, and night conditions. (Chapter 6 discusses the third phase, **advanced gunnery**.)*

***WARNING**

Before allowing anyone to move between stations, ensure that--

- **All rifles and grenade launchers have been cleared.**
- **Bolts are to the rear.**
- **Barrel assemblies are in the open position.**

Anyone observing an unsafe act should call out "Cease fire" and notify range personnel immediately.

Section I. PRELIMINARY MARKSMANSHIP TRAINING

Grenadiers and leaders must master marksmanship fundamentals before firing individually or collectively. During preliminary marksmanship training, grenadiers learn and demonstrate the individual skills that prepare them to fire live ammunition. After learning the characteristics and mechanics of the weapon (Chapters 2, 3, and 4), they learn the four fundamentals of marksmanship, sight manipulation, and response to fire commands. Dry-fire exercises are excellent for training to proficiency. Good preliminary marksmanship instruction improves individual proficiency, which in turn improves the proficiency of collective fire.

*5-1. FOUR FUNDAMENTALS OF MARKSMANSHIP

The four fundamentals of M203 marksmanship are steady position, aiming, breathing, and trigger control. Only the first fundamental (steady position) varies. The other three remain the same regardless of the Soldier's position.

a. **Steady Position**. This varies according to the position and the type of sight used (quadrant or leaf). Note that for firing prone, a supported position is best.

(1) ***Prone Position, Quadrant Sight*** (top, Figure 5-1).

(a) Lie face down, grasp the M16 pistol magazine with your right hand, and place the butt of the rifle into the pocket of your right shoulder.

(b) Lower your right elbow to the ground so your shoulders are level. This places the weight of your body behind the weapon, which enables you to recover quickly each time you fire.

(c) Grasp the barrel grip with your left hand, supporting with sandbags. Straighten your upper body and spread your legs a comfortable distance apart. Try to point your toes outward and relax your ankles so your heels will rest on the ground. Relax the weight of your upper body forward onto your left arm.

(2) ***Prone Position, Leaf Sight*** (bottom, Figure 5-1). While firing with the leaf sight at ranges greater than 150 meters, place the butt stock of the weapon under your armpit and grip firmly to prevent the weapon from moving. Lean your head 45 degrees to the right and place the M16 front sight post on the desired range. Raise the butt stock and lower the muzzle to obtain the proper sight alignment and sight picture.

Figure 5-1. Prone supported position, quadrant sight (top) and leaf sight (bottom).

***WARNING**

Before allowing anyone to move between stations, ensure that the sling is clear of the weapon muzzle before firing.

(3) ***Kneeling Position, Quadrant Sight*** (left, Figure 5-2).

(a) Kneel on your right knee while facing the target with your right hand on the magazine and your left hand grasping the barrel grip.

(b) Place your left foot about .45 meter (18 inches) to your left front with your toes pointing in the general direction of the target.

(c) Keeping your right toe in place, sit on your right heel.

(d) Place your left elbow forward of your left knee, resting the flat portion of your upper arm on your knee.

(e) Move the rifle butt into the pocket of your right shoulder, pulling the rifle magazine with your right hand and grasping the barrel grip with your left hand.

(f) With your right hand on the rifle magazine, place your right forefinger in the trigger guard of the grenade launcher.

(g) Pull the rifle firmly into your shoulder.

(h) Pull your right elbow in close to your body to help you apply rearward pressure to the weapon. Ensure that your leg completes a solid, three-point base for your position.

(4) ***Kneeling Position, Leaf Sight*** (right, Figure 5-2).

(a) For ranges greater than 150 meters, place the butt stock of the weapon under your armpit and grip firmly to prevent the weapon from moving.

(b) Lean your head 45 degrees to the right and place the front sight post of the M16 on the desired range. Raise the butt stock and lower the muzzle to obtain the proper sight alignment and sight picture.

Figure 5-2. Kneeling position, quadrant sight (L) and leaf sight (R).

(5) ***Sitting Position, Open-Legged, Quadrant Sight*** (left, Figure 5-3, page 5-4).

(a) Sit down, breaking your fall with your right hand, and slide your buttocks well to the rear. Face the target half right, and spread your feet wide.

(b) Grasp the rifle magazine with your right hand and the barrel grip with your left hand.

(c) Bend forward from your hips.

(d) Move the butt of the rifle into the pocket of your right shoulder, still holding the rifle magazine with your right hand.

(e) Pull the weapon down slightly with your left hand and pull it to the rear firmly with your right hand.

(6) ***Sitting Position, Open-Legged, Lea f Sight*** (right, Figure 5-3).

(a) For ranges greater than 150 meters, place the butt stock of the weapon under your armpit and grip firmly to prevent the weapon from moving.

(b) Lean your head 45 degrees to the right and place the M16 front sight post on the desired range. Raise the butt and lower the muzzle to obtain the proper sight alignment and sight picture.

Figure 5-3. Sitting position, open-legged, quadrant sight (L) and leaf sight (R).

(7) ***Sitting Position, Cross-Ankle, Quadrant Sight*** (left, Figure 5-4).

(a) Sit facing the target half right.

(b) Extend your legs from your body and cross your left ankle over your right ankle.

(c) Keep both ankles straight.

(d) Grasp the rifle magazine with your right hand and the barrel grip with your left.

(e) Place your left upper arm across your left knee.

(f) Move the butt of the rifle into the pocket of your right shoulder.

(8) ***Sitting Position, Cross-Ankle, Leaf Sight*** (right, Figure 5-4).

(a) For ranges greater than 150 meters, place the butt stock of the weapon under your armpit and grip firmly to prevent the weapon from moving.

(a) Lean your head 45 degrees to the right and place the M16 front sight post on the desired range. Raise the butt and lower the muzzle to obtain the proper sight alignment and sight picture.

Figure 5-4. Sitting position, cross-ankled, quadrant sight (L) and leaf sight (R).

(9) ***Sitting Position, Cross-Legged, Quadrant Sight*** (left, Figure 5-5).

(a) Sit down facing the target half right.

(b) Cross your left leg over your right leg and draw both feet close to your body.

(c) Grasp the rifle magazine with your right hand.

(d) Move the butt of the rifle into the pocket of your right shoulder, and grasp the rifle barrel grip properly with your left hand.

(10) ***Sitting Position, Cross-Legged, Leaf Sight*** (right, Figure 5-5).

(a) For ranges greater than 150 meters, place the butt stock of the weapon under your armpit and grip firmly to prevent the weapon from moving.

(b) Lean your head 45 degrees to the right and place the M16 front sight post on the desired range. Raise the butt and lower the muzzle to obtain the proper sight alignment and sight picture.

Figure 5-5. Sitting position, cross-legged, quadrant sight (L) and leaf sight (R).

(11) ***Squatting Position, Cross-Legged, Quadrant Sight*** (left, Figure 5-6, page 5-6).

(a) Turn half right to the target and, keeping both feet flat on the ground and a comfortable distance apart, squat as low as you can.

(b) Grasp the rifle magazine with your right hand.

(c) Place your left upper arm inside your left knee and the butt of the rifle into the pocket of your right shoulder. Grasp the rifle barrel grip properly.

(d) Lower your right elbow against the inside of your right knee.

(12) ***Squatting Position, Cross-Legged, Leaf Sight*** (right, Figure 5-6, page 5-6).

(a) For ranges greater than 150 meters, place the butt stock of the weapon under your armpit and grip firmly to prevent the weapon from moving.

(b) Lean your head 45 degrees to the right and place the M16 front sight post on the desired range. Raise the butt and lower the muzzle to obtain proper sight alignment and sight picture.

Figure 5-6. Squatting position, quadrant sight (L) and leaf sight (R).

(13) ***Fighting Position, Quadrant Sight*** (left, Figure 5-7).

(a) If possible, use support when firing from a fighting position.

(b) Place your right foot against the rear of the fighting position and lean forward until your chest is against its forward edge.

(c) Grasp the magazine with your right hand.

(d) Place your left elbow on or against solid support.

(e) Use your right hand to position the butt of the rifle in the pocket of your right shoulder. Grasp the rifle barrel grip properly.

(f) Place your right elbow on or against a solid support and relax into a comfortable firing position.

(14) ***Fighting Position, Leaf Sight*** (right, Figure 5-7).

(a) For ranges greater than 150 meters, place the butt stock of the weapon under your armpit and grip firmly to prevent the weapon from moving.

(b) Lean your head 45 degrees to the right and place the M16 front sight post on the desired range. Raise the butt and lower the muzzle to obtain proper sight alignment and sight picture.

Figure 5-7. Fighting position, quadrant sight (L) and leaf sight (R).

NOTE: The weapon must not touch the support.

(15) ***Standing Position, Quadrant Sight*** (left, Figure 5-8).

(a) Face the target while standing with your feet spread a comfortable distance apart.

(b) Grasp the rifle barrel grip with your left hand and the rifle magazine with your right hand.

(c) Place the butt of the stock into your right shoulder so that the sight is level with your eyes.

(d) Hold your right elbow high to form a good pocket for the butt of the stock and to permit a strong rearward pressure with your right hand.

(e) Hold most of the weight of the weapon with your left hand.

(f) Shift your feet until you achieve a natural aiming stance.

(16) ***Standing Position, Leaf Sight*** (right, Figure 5-8).

(a) For ranges greater than 150 meters, place the butt stock of the weapon under your armpit and grip firmly to prevent the weapon from moving.

(b) Lean your head 45 degrees to the right and place the M16 front sight post on the desired range. Raise the butt and lower the muzzle to obtain the proper sight alignment and sight picture.

Figure 5-8. Standing position, quadrant sight (L) and leaf sight (R).

b. **Aiming**. Aiming procedures for every position follow:

(1) ***Aligning Sight***. When using the leaf sight, align it with the front sight post of the M16. When using the quadrant sight, align its rear sight aperture with its front sight post. Picture a horizontal line through the center of the leaf sight or rear sight aperture: the top of the M16's front sight post should touch this line. Picture a vertical line through the center of the leaf sight or rear sight aperture: this line should vertically bisect the front sight post (Figure 5-9, page 5-8).

(2) ***Focusing***. For either sight, focus on the front sight post. A good firing position places your eye directly on line with the center of the leaf sight or rear sight aperture. Your eye's natural ability to center objects in a circle and seek the point of greatest light will help you align the sight correctly.

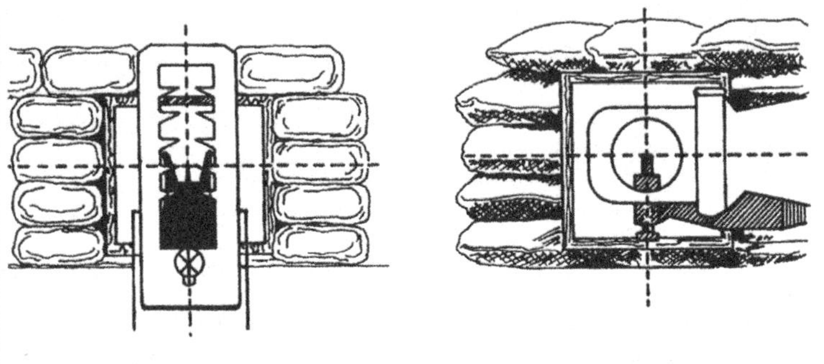

Figure 5-9. Sight pictures for leaf and quadrant sights.

(3) ***Obtaining Sight Picture***. To achieve a correct sight picture, align the front sight post and the leaf sight or rear sight aperture with the target. For area targets, aim where the round's bursting radius will make the round most effective. For point targets, aim at the target's center of mass.

c. **Breathing**. The technique for breathing is the same for every position: Breathe naturally, exhale most of your air, hold your breath, and fire before you become uncomfortable. In combat, just choke off your breath before firing.

d. **Trigger Control**. The technique for trigger control is the same for every position. Place your trigger finger (the index finger of your right hand) so that the trigger is between the first joint and the tip of your finger (not at the extreme end of your finger). Adjust for your hand size and grip. Then, squeeze your trigger finger to the rear without disturbing the lay of the weapon.

5-2. LIMITED VISIBILITY

The fundamentals of marksmanship are almost the same in limited visibility as in normal visibility.

a. **Steady Position**. An M203 with an AN/PVS-4 mounted on it leans to the left. When assuming a steady position, the grenadier must apply more rearward pressure to compensate for the lean and then steady the weapon.

b. **Aiming**. The grenadier sights with the reticle of the AN/PVS-4 rather than with the M203's iron sights. Sighting this way requires him to change position, which breaks his stock weld and makes the weapon seem heavier.

c. **Breathing**. Though breathing itself is affected little by limited visibility, using night vision devices that magnify the field of view increases the effect of weapon movement caused by breathing.

d. **Trigger Control**. This is the same regardless of visibility conditions. The objective is to keep the weapon aligned with the target.

*e. **Night Vision Devices**. The AN/PVS-7 is issued for use with the M203, whereas the AN/PVS-4 is normally issued for use with crew-served weapons. M203 gunners may qualify with either device. In a defensive position, the gunner identifies targets during daylight and constructs aiming or elevation stakes. Because the AN/PVS-7 rear sight must be set to the far setting to sense rounds, the gunner cannot see both the M203 sights and the target at the same time. Therefore, stakes are more important with the AN/PVS-7 than with the AN/PVS-4.

f. **Marked-Sling Method**. The best field-expedient method for firing the M203 grenade launcher in limited visibility is the marked-sling method using only the M16 rifle series (Figure 5-10, page 5-10). To use this method, the grenadier must--

(1) Face the target and kneel on the right knee (if firing right-handed), keeping the left foot pointed toward the target.

(2) Loosen the sling and place the forward foot in the sling.

(3) Place the butt of the stock firmly on the ground.

(4) Using the left hand, grasp firmly the upper barrel grip just below the barrel.

(5) Grasp the receiver group with the right hand.

WARNING

Placing your knee against the buttstock could injure your knee.

(6) Ensure the sling is taut and vertical between the front sling swivel and the boot (Figure 5-11, page 5-10). If not, the rounds will impact at a greater range than desired.

(7) Fire several rounds to determine the desired range.

(8) Mark the sling (where it is held to the ground by the foot) with colored tape, paint, ink, or whatever is available. Mark the position of the buckles (Figure 5-12, page 5-10) so that, if either is moved, the grenadier can return them to their original positions and be assured of constant range accuracy.

(9) If the sling gets wet, it may stretch or shrink, indirectly causing the rounds to impact closer or farther than desired.

*5-3. CHEMICAL, BIOLOGICAL, RADIOLOGICAL, AND NUCLEAR ENVIRONMENTS

The fundamentals of marksmanship remain valid in the CBRN environment, but some modifications may be needed to accommodate the equipment.

a. **Steady Position**. Bulky CBRN wear requires the grenadier to press the stock of the weapon more firmly into his shoulder pocket.

b. **Aiming**. Aiming is affected little by CBRN.

c. **Breathing**. Wearing the protective mask makes breathing more difficult. Grenadiers must try to breathe normally to avoid hyperventilating while firing.

d. **Trigger Control**. All Soldiers must wear rubber gloves.

Figure 5-10. Marked-sling method.

Figure 5-11. Front-sling swivel and front of boot.

Figure 5-12. Marked sling and buckle highlighted.

5-4. FIRE COMMANDS

Standard fire commands are explained to grenadiers and are used during all subsequent gunnery training. Trainers give the appropriate elements before each dry-fire or live-fire exercise. The grenadier performs as directed and repeats each element as it is announced. (Chapter 6 discusses fire commands in detail.)

a. **Alert**. The trainer gives the alert as a fire mission. On hearing this, the grenadier loads the weapon and moves the safety lever to "F."

b. **Direction**. The trainer gives the direction to target.

c. **Description**. The trainer describes the target, for example, "Bunker" or "Machine gun position," and the grenadier lays on the target.

d. **Range**. The trainer gives the (estimated) range to the target, for example, "150."

e. **Method of Fire**. The method of fire for either target is three rounds. On the basic range, grenadiers fire at both point and area targets.

f. **Command to Open Fire**. To open fire, the trainer commands COMMENCE FIRING or AT MY COMMAND. When ready, the grenadier announces "Up" and fires or waits for the command to fire. When all grenadiers are ready, the trainer gives the actual command to fire.

5-5. DRY-FIRE EXERCISES

Dry-fire exercises train grenadiers in the techniques of loading, unloading, immediate action, fundamentals of marksmanship, and sight manipulation. These exercises are conducted with TP or dummy rounds. The trainer gives fire commands as appropriate.

a. **Loading and Unloading Exercise**. This trains the grenadier to operate and clear the weapon proficiently. Loading and unloading procedures (Chapter 3) should be practiced with dummy ammunition.

b. **Immediate Action Exercise**. This exercise is conducted with a dummy round and the basic grenade launcher target.

(1) Load the weapon with a dummy round and aim it at one of the targets on the basic grenade launcher range.

(2) Maintain the sight picture while you pull the trigger to simulate firing.

(3) When you are informed that you have a misfire, apply misfire procedures; then continue to fire (Chapter 4).

c. **Aiming Exercise**. This exercise requires the grenadier to simulate firing a dummy round at a target on the basic grenade launcher range.

(1) Maintain your sight picture throughout the firing cycle.

(2) If, after firing, you note that the sight picture has moved, then you were unsteady when you fired.

(3) After each shot, apply immediate action procedures to extract and eject the dummy cartridges. Then recock the barrel assembly.

d. **Sight Setting and Sight Changing Exercises**. These exercises train the grenadier to operate and adjust both quadrant and leaf sights.

(1) ***Range***. Manipulate the sights to different range settings (quadrant sight, 50 to 400 meters; leaf sight, 50 to 250 meters). To learn to make fine adjustments for elevation, manipulate the sights from the minimum to the maximum setting. When you do not have time to adjust the sights, you may adjust the aiming point instead.

(2) ***Windage***. Depress the rear sight aperture left and right and traverse the windage screw across the entire scale.

e. **Dry-Fire Proficiency (Performance) Exam**. Grenadiers practice the dry-fire tasks until they become proficient in operating the weapon; then they take the dry-fire proficiency exam (Appendix C). This exam emphasizes learning by doing. Before he can progress to live firing, each grenadier must demonstrate skill in every task in the exam.

f. **Remedial Training**. Soldiers who fail to pass the performance exam must attend remedial training, after which they retest. Those who pass can help train those having difficulty.

5-6. SENSING AND ADJUSTMENT OF FIRE

The grenadier determines (senses) where the grenade landed relative to the target and then adjusts elevation and deflection.

a. **Sensing**. As soon as the grenade explodes, determine where it exploded with respect to the target. This is called "sensing" (the impact) and has two aspects: range and deviation. Because the casualty radius of the HE round is 5 meters (5 1/2 yards), determine both range and deviation to the nearest 5 meters.

(1) ***Range***. Sense the range as one of the following:

(a) *Short*. The grenade bursts between you and the target.

(b) *Over*. The grenade bursts beyond the target.

(c) *Target*. The grenade hits any part of the target.

(d) *Range Correct*. The grenade bursts slightly left or right of the target, but at the correct range.

(e) *Doubtful*. The grenade burst left or right of the grenadier, but you cannot sense the range.

(2) ***Deviation***. Announce a deviation sensing as either--

- Right or left of the target, or
- On line with the target.

b. **Adjustment of Fire**. To ensure a second-round hit, adjust your fire by sensing the impact of the round and manipulating the sight.

(1) If time allows, whether using the AN/PVS-4 or AN/PVS-7, adjust the sights; if time is critical, adjust the point of aim instead.

(2) If the grenade lands more than 25 meters over or short of the target, adjust the range quadrant to bring the next grenade on target.

(3) If the grenade explodes less than 25 meters from the target, adjust the point of aim to bring the next grenade on target.

(4) If the launcher is properly zeroed, deviation errors are normally small and easily corrected by adjusting the aiming point. A wind strong enough to move the grenade out of its normal trajectory, however, increases the size of the deviation errors. After observing the effect of the wind on the strike of the grenade, compensate for the effect of the wind by aiming into it. This should help bring the next grenade on target. For example, if the grenade bursts to the left and short of the target, sense the strike of the round relative to the target and then adjust an equivalent distance to the right and over the target to achieve a target hit. Watch the flight of the grenade to the target. This helps determine the effect of the wind on the grenade as it moves toward the target. Evaluating and compensating for the wind before firing increases your chances of achieving a first-round hit.

*5-7. LAYOUT OF GRENADE LAUNCHER RANGE

The grenade launcher range is designed for all grenade launchers. Because Soldiers can qualify on this range in all conditions, it prepares grenadiers for combat situations. The range has four stations (Figure 5-13). Minimum range personnel and their duties are the same for M203 qualification firing as they are for other grenadier firing. These personnel include an OIC, NCOIC, safety officer, ammunition NCO, tower operator, station NCOs, primary trainer, and concurrent training trainers. However, local policy may require more personnel. (Appendix D discusses range safety; TC 25-8 provides a detailed setup and target configuration for this range.)

5-8. DESCRIPTION OF RANGE AND TARGETS

The range has four self-contained stations (Figure 5-13). It is 30 meters wide by 500 meters deep and has a no-HE fire zone out to 165 meters. (Grenadiers can fire HE only on Stations 1, 3, and 4.) Targets should be built from durable materials to reduce downrange target maintenance. Those within each station must be grouped and spaced so that the grenadier may fire on close-range, mid-range, and long-range targets, in that order. The following description of the stations and targets is included to help trainers maintain control during zeroing, practice, and record fire. Figure 5-14 shows the range in meters and feet for combat, point target, training, area target and maximum ranges for the M203.

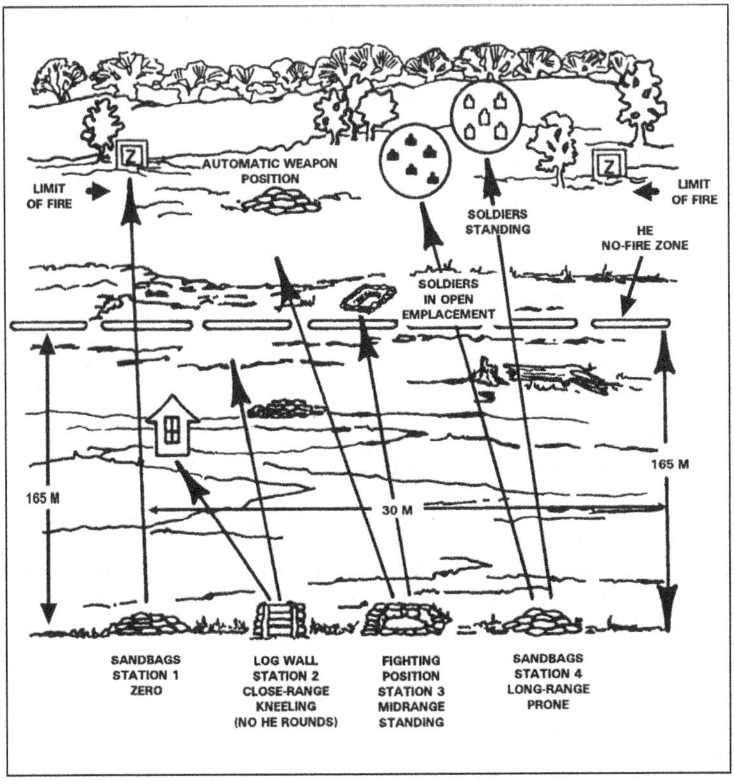

Figure 5-13. Grenade launcher range.

WARNING

Before firing HE grenades, move to a protected (covered) position. The minimum safe ranges follow:

Training	165 meters	(541 feet)
Combat	31 meters	(102 feet)

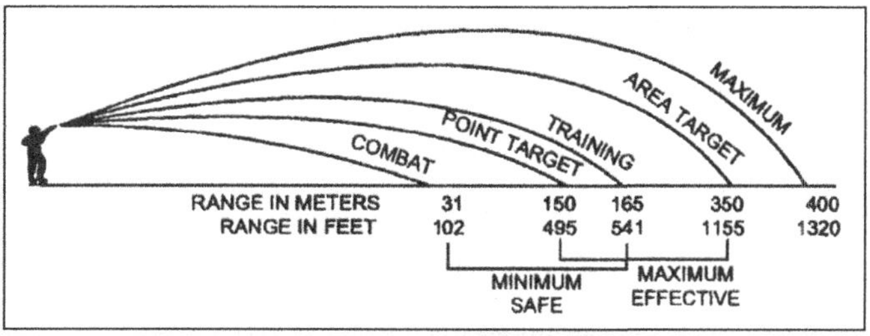

Figure 5-14. Ranges in meters and feet.

NOTE: To simplify the task of finding and destroying duds, trainers must ensure the impact area is free of any vegetation other than short grass.

a. Station 1 consists of a prone fighting position with a log or sandbag support and a zeroing target at 200 meters. The target should be constructed of logs or other suitable material. It must have a surface at least 2 meters high by 2 meters wide (6 feet by 6 feet). The target should be clearly marked with a large "Z" painted in a color that contrasts with the surrounding background and that is visible in different sun or glare conditions.

b. Station 2 consists of an upright log or log wall, a kneeling firing position about 4 feet high, and 2 point-type targets. The targets include a simulated window or door of a building at 90 to 100 meters and a small bunker or fighting position with overhead cover at 125 meters. The targets may be constructed of logs, sandbags, or other suitable material.

c. Station 3 consists of a fighting position and two targets. The targets are a two-person bunker at 150 meters and an automatic weapon position at 175 meters. The bunker represents a point target, and the automatic weapon position represents a target that can be engaged with area-type fire. The targets may be constructed of logs, sandbags, or other suitable material.

d. Station 4 consists of a prone fighting position with a log or sandbag support and two area-type targets (with personnel targets in the open) at 250 and 350 meters. The log or sandbags at the firing position are used for support and cover. The targets are E-type and F-type shaped silhouette and are made out of durable materials.

Section II. BASIC GUNNERY

Basic gunnery allows the grenadier to zero and apply the fundamentals of marksmanship during live-fire exercises in day, night, and CBRN conditions.

5-9. ZEROING OF M203 GRENADE LAUNCHER

A correct zero consists of the elevation and windage sight settings that enable the grenadier to hit the point of aim at a given range with one of the three sighting systems: leaf, quadrant, or night sight. To zero the M203 using either the leaf sight or quadrant sight, the grenadier engages a target at 200 meters. (The M203 is normally zeroed using only the quadrant sight, but may be zeroed with both sights or with only the leaf sight.)

a. **Zeroing the Leaf Sight**. A red mark at 50 meters on the leaf sight reminds the grenadier not to zero at this range.

(1) Select a target at 200 meters.

(2) Place the sight in the upright position.

(3) Place the center mark of the windage scale on the index line on the rear of the sight base.

(4) Loosen the elevation adjustment screw on the leaf sight.

(5) Place the leaf sight's index line on the sight mount's center elevation mark.

(6) Tighten the elevation adjustment screw.

(7) Assume a prone supported firing position.

(8) Load one round of 40-mm HE or TP ammunition.

(9) Use correct sighting and aiming procedures to align the target with the front leaf sight.

(10) Fire a round, sense the impact, and adjust the sight.

(a) *Windage*. Turn the sight windage screw clockwise to move the leaf sight to the left and vice versa. One increment moves round impact 1 1/2 meters at a range of 200 meters.

(b) *Range*. Use a 40-mm cartridge case and turn the elevation adjustment screw to raise the leaf sight (this increases range) or to lower the leaf sight (this decreases range). Moving the leaf sight one increment moves the impact of the round 10 meters at a range of 200 meters.

(11) Fire two more cartridges, readjusting the sight after each. Once a round impacts within 5 meters of the target, the weapon is zeroed.

(12) After you have zeroed the weapon, record the zero data on your scorecard. As soon as you can, transfer the information to a small piece of paper and tape this inside the M16 pistol grip.

b. **Zeroing the Quadrant Sight**.

(1) Select a target at 200 meters.

(2) Ensure that the quadrant sight is correctly mounted on the rifle's carrying handle.

(3) Open the front sight post and rear sight aperture.

(a) Move the front sight post to its highest position and then back 2 1/2 turns.

(b) Depress the rear sight retainer. Slide the rear sight aperture to the left until its white index line aligns with the edge of the sight aperture arm.

(4) Move the sight latch rearward and reposition the quadrant sight arm to zeroing range (200 meters).

(5) Assume a prone supported firing position.

(6) Use correct sighting and aiming procedures to align the target with the front sight post and rear sight aperture.

(7) Load one round of 40-mm HE or TP ammunition.

(8) Fire a round, sense the impact, and adjust the sight.

(a) *Elevation*. Turn the front sight post right (to decrease elevation) or left (to increase elevation). At a range of 200 meters, one full turn equals 5 meters.

(b) *Windage*. Press the sight aperture retainer. Move the rear sight aperture away from the barrel to move the trajectory to the left or vice versa. At a range of 200 meters, one notch on the rear sight aperture equals 1 1/2 meters.

(9) Fire two more cartridges, readjusting the sights after each. If the round lands within 5 meters of the target, the weapon is zeroed.

(10) After you have zeroed the weapon, record the zero data on your scorecard. As soon as you can, transfer this information to a small piece of paper and tape this inside the M16 pistol grip.

5-10. OVERALL QUALIFICATION STANDARDS

DA Form 2946-R (40-mm Grenade Launcher Scorecard) is used for qualification firing. To qualify with an M203, a grenadier must perform to prescribed standards and must score at least 60 of 90 possible points. Each target hit is worth 10 points. Zeroing is not included on the scorecard because the weapon must be zeroed before qualification firing. However, the zero data should already have been entered on the scorecard when the weapon was zeroed. HE familiarization may be included in qualification firing but is not scored. Ratings are awarded based on the point chart shown on the scorecard. A blank scorecard is provided in back of this field manual and must be locally reproduced on 8 1/2- by 11-inch paper. Figure 5-15, page 5-18 shows an example of a completed scorecard.

*5-11. DAY RECORD FIRE

Day record fire gives the grenadier the confidence and experience he needs to progress from dry-fire exercises to record fire. All Soldiers must be prepared to accomplish their missions, even in protective clothing; thus, day record fire includes two CBRN tasks (Tasks 4 and 5). This exercise is conducted on a grenade launcher range IAW Firing Table I (Table 5-1, page 5-20). Before they fire for qualification, grenadiers must first zero their weapons and receive instruction on the objectives, range, targets, and qualification standards. The unit is organized in firing orders based on range constraints. Each firing order consists of two grenadiers, one of whom assists. Grenadiers fire this exercise from the following fighting positions: kneeling supported, mid-range supported, long-range supported, CBRN mid-range point target, and CBRN mid-range area target. For each of these tasks, the grenadier can designate which target he will engage first.

40-MM GRENADE LAUNCHER SCORECARD
For use of this form, see FM 3-22.31; the proponent agency is TRADOC.

EXAMPLE

1. ID CODE *	2. UNIT	3. DATE *(YYYYMMDD)*
20070419BTD01	C CO, 2 29 IN	20070419

4. ZERO LEAF SIGHT	DEFILADE	ELEVATION	5. ZERO QUADRANT SIGHT	DEFILADE	ELEVATION
				L4	R1

6. RECORD FIRE	TASK NO.	TIME (MIN)	TGT	HIT	MISS
DAY	1	2	1	X	
			2		X
	2	2	1	X	
			2	X	
	3	2	1	X	
			2	X	
CBRN	4	2	1	X	
	5	2	1	X	
NIGHT	6	2	1	X	
* Do not use personal information	7. TOTAL POINTS *(Award 10 points for each hit)*			8	

8. RATING SCALE

TOTAL POINTS *(Block 7)*	RATING	
80 - 90	EXPERT	☒
70 - 75	FIRST CLASS	☐
60 - 65	SECOND CLASS	☐
0 - 55	UNQUALIFIED	☐

9. GRADER'S INITIALS	10. DATE *(YYYYMMDD)*
RM	20070419
11. OIC'S INITIALS	**12. DATE *(YYYYMMDD)***
CS	20070419

DA FORM 2946-R, FEB 2007 EDITION OF NOV 2002 IS OBSOLETE. APD V1.00

Figure 5-15. Example completed DA Form 2946-R, 40-mm Grenade Launcher Scorecard.

WARNING

Before allowing anyone to move between stations, ensure that--

- **All weapons are clear.**
- **Bolts are to the rear.**
- **Barrel assemblies are open.**

If you see an unsafe act, call out "Cease fire" and notify range personnel immediately.

a. **Station 1, Zeroing**. The grenadier zeroes with both quadrant and leaf sights at Station 1.

(1) ***Leaf Sight***. From a prone supported firing position, fire to zero the weapon. This reinforces the experience gained during dry firing and allows practice in loading and firing with the most accurate sensing and adjustments obtainable. If you zero in three rounds, use the other two rounds to confirm the zero. If you cannot zero with five rounds, the trainer must remove you from the firing line for remedial training.

(a) Prepare the sight for zeroing.

(b) Assume a good prone supported firing position.

(c) When you receive the following fire command, repeat each element as it is given:

GRENADIER
FRONT
200 (ZERO PANEL)
ONE ROUND
COMMENCE FIRING

(d) Load one round, obtain the proper sight picture, and announce "Up" to your assistant.

(e) When the tower operator gives the command to commence firing, fire one round at the panel marked "Z."

(f) Sense the impact of the round. If the round did not land within 5 meters of the zero panel, adjust the sights for windage and elevation.

(g) Repeat until a round lands within 5 meters of the zero panel.

(h) Once you have zeroed the weapon, record the zero data on your scorecard. As soon as you can, transfer the information to a small piece of paper and tape this inside the M16 pistol grip.

(2) ***Quadrant Sight***. From a prone supported firing position, fire to zero the weapon. This reinforces the experience gained during dry firing and gives you practice in loading and firing with the most accurate sensing and adjustments you can obtain. The steps for zeroing with the quadrant sight are the same as those for zeroing with the leaf sight.

b. **Station 2, Task 1, Kneeling Supported Position**. (Only TP rounds may be used at this station.)

(1) When you receive the command DESIGNATE THE TARGET, identify the target you intend to engage by announcing WINDOW or BUNKER.

(2) When you receive the command DETERMINE THE RANGE, announce the range to the target.

(3) Load one of the three rounds allotted. Because HE must not be fired at ranges of less than 165 meters on the basic grenade launcher range, use only TP rounds.

(4) When you receive the following fire command, repeat each element as it is given:

GRENADIER
FRONT
3 ROUNDS
90 meters (window) or 100 meters (bunker)
COMMENCE FIRING

(5) Acquire the proper sight picture and announce "Up" to the grader.

(6) Engage the target given in the fire command until you hit it. Fire any remaining rounds at the second target. You need no other fire command. For each round you fire, your assistant announces "Hit" or "Miss."

STATION	TASK	TIME	ROUNDS	TYPE	TARGET(S) AND RANGE(S)
2	1	2 min	3	TP	Window at 90 to 100 meters Bunker at 105 to 125 meters
3	2	2 min	3	HE	Bunker at 170 to 180 meters Automatic weapon at 190 to 200 meters
4	3	2 min	3	HE	Troops in open emplacement at 275 to 300 meters Troops in open at 235 to 350 meters
3	4 CBRN	2 min	3	HE	Bunker at 170 to 180 meters
3	5 CBRN	2 min	3	HE	Automatic weapons at 190 to 200 meters

Table 5-1. Firing Table I, day record fire qualification.

c. **Station 3, Task 2, Midrange Supported Position.**

(1) When you receive the command DESIGNATE THE TARGET, identify the target you intend to engage by announcing "Bunker" or "automatic weapon."

(2) When you receive the command DETERMINE THE RANGE, announce the range to the target.

(3) Load one of the three rounds allotted.

(4) When you receive the following fire command, repeat each element as it is given:

GRENADIER
FRONT
3 ROUNDS
170 METERS (BUNKER) OR 190 METERS (AUTOMATIC WEAPON)
COMMENCE FIRING

(5) Acquire the proper sight picture and announce "Up" to the grader.

(6) Engage the target given in the fire command until you hit it. Fire any remaining rounds at the second target. You need no other fire command. For each round you fire, your assistant announces "Hit" or "Miss."

d. **Station 4, Task 3, Long-Range Supported Position.**

(1) When you receive the command DESIGNATE THE TARGET, identify the target you intend to engage by announcing "Troops in the open emplacement" or "Troops in the open."

(2) When you receive the command DETERMINE THE RANGE, announce the range to the target.

(3) Load one of the three rounds allotted.

(4) When you receive the following fire command, repeat each element as it is given:

GRENADIER
FRONT

3 ROUNDS
300 METERS (TROOPS IN THE OPEN EMPLACEMENT) or
350 METERS TROOPS IN OPEN
COMMENCE FIRING

(5) Acquire the proper sight picture, and announce "Up" to the grader.

(6) When the tower operator gives the command FIRE, engage the target given in the fire command until you hit it. Fire any remaining rounds at the second target. You need no other fire command. Before firing, you must know the procedure to follow in the event of a stoppage. For each round you fire, your assistant announces "Hit" or "Miss."

e. **Station 3, Task 4, Midrange Supported Position, Point Target, CBRN**.

(1) Put on, clear, and check your mask within 9 seconds. Within the next 6 seconds, pull the hood over your head and zip the front of it closed.

(2) Load one of the three rounds allotted.

(3) When you receive the following fire command, repeat each element as it is given:

FIRE MISSION
FRONT
3 ROUNDS
170 METERS (BUNKER)
AT MY COMMAND

(4) Acquire the proper sight picture and announce "Up" to your assistant.

(5) Have your assistant signal the tower operator that you are ready.

(6) When the tower operator gives the command to commence firing, engage the target given in the fire command until you hit it. For each round you fire, your assistant announces "Hit" or "Miss."

f. **Station 3, Task 5, Midrange Supported Position, Area Target, CBRN**.

(1) Load one of the three rounds allotted.

(2) When you receive the following fire command, repeat each element as it is given:

FIRE MISSION
FRONT
3 ROUNDS
190 METERS (AUTOMATIC WEAPON POSITION)
AT MY COMMAND

(3) Acquire the proper sight picture and announce "Up" to your assistant.

(4) Have your assistant signal the tower operator that you are ready.

(5) When the tower operator gives the command to FIRE, engage the target given in the fire command until you hit it. For each round you fire, your assistant announces "Hit" or "Miss."

5-12. DAY RECORD FIRE QUALIFICATION STANDARDS

Before qualification firing, each grenadier must know the tasks, the time and ammunition required, the procedures to follow if a stoppage occurs, the penalties for failure to stop firing when commanded or signaled to do so, and the method used for scoring targets.

a. **Time and Ammunition**. Each grenadier determines the target and its distance before loading any rounds. When the grenadier receives the command to FIRE, the time allotted for that task in Firing Table I begins.

b. **Stoppages**. The grenadier must apply immediate action procedures if a stoppage occurs. If he can reduce the stoppage, he can continue to fire the course. The trainers allow the grenadier an extra 15 seconds for each application of immediate action.

(1) If a stoppage occurs that you cannot reduce by immediate action, raise your hand and announce "Time."

(2) When you say "Time," the assistant trainer notes the time, ensures that a real stoppage exists, and tries to clear the stoppage. If he clears it, you can complete firing. If he cannot clear it, the grader will clear it, and you will be allowed 15 seconds for each round remaining to complete firing.

(3) If you made an error that caused the stoppage, you do not receive extra time, and your score consists only of whatever you had earned when the stoppage occurred.

(4) If the grenade launcher must be replaced, you are allotted 10 rounds to zero a new one; then you may repeat the exercise.

(5) If malfunctions prevent you from finishing the exercise in the time allowed, you can finish it in an "alibi run" after all other grenadiers complete firing.

c. **Penalties**. Five points are deducted from the score of any grenadier who fails to stop firing when the trainer commands or signals to do so. If a grenadier fires at the wrong target, he loses the rounds allotted for the other target, which leaves him only the remainder of his rounds to expend on both targets.

d. **Target Scoring**. The trainer or assistant trainer records scores on DA Form 2946-R. They determine whether each grenade fired is a hit or miss and assign 0 points for a miss or 10 points for a hit. Tasks 1 through 3 each consist of two targets, so the total available for each of these tasks is 20 points. The grenadier may select which of the two targets to engage first. If he scores a hit on the first, the trainer permits him to engage the second. Once he hits both targets, he returns any unexpended rounds to the assistant trainer. Tasks 4 and 5 each consist of firing one target for a total of 10 points each.

(1) ***Window or Door***. To score a hit on a window or door, the grenade must either strike the target or go through the opening in the center of the target.

(2) ***Bunker***. To score a hit on a bunker, the grenade must strike anywhere on the face of the bunker.

(3) ***Automatic Weapon***. To score a hit on an automatic weapon, the grenade must strike within 5 meters of the target.

(4) ***Troops***. To score a hit on troops, the grenade must strike within 5 meters of the target.

5-13. MOUNTING OF AN/PVS-4 (WITHOUT THE RAIL SYSTEM)

The grenadier must mount the AN/PVS-4 to the weapon before he zeroes it, and he must do both before he can qualify with the M203 grenade launcher. To mount the scope, the grenadier must--

a. Remove the quadrant sight.

b. Position the mounting bracket assembly on the left side of the rifle so that the two clamps project through the opening under the handle. Loosen the wing nuts completely (Figure 5-16, page 5-24).

c. Turn the clamp plates so that the pointed ends are in the UP position and are seated against the handle.

d. Tighten the wing nuts clockwise until the mounting bracket is secure against the weapon.

e. Position the sight in the groove on top of the bracket and align the threaded hole in the base of the sight mounting adapter with the lever screw assembly. Tighten the screw clockwise firmly to secure the sight to the bracket.

5-14. ZEROING OF AN/PVS-4 TO M203

After being mounted on the M203, the nightsight must be zeroed to the M16 before it can be zeroed to the M203. FM 3-22.9 provides instructions for doing this. The M16 is then used to zero the nightsight to the M203. The grenade launcher rounds are fired only to confirm the zero. To zero the nightsight to the M203--

a. **Center the Reticle Pattern**. Use the aiming points on the nightsight reticle (Figure 5-17, page 5-24) and the range settings on the mounting bracket. Center the nightsight's reticle pattern within the field of view (FOV). Note that it may not be centered even if it appears to be. To ensure it is, rotate the azimuth control knob either way until it stops. Then, rotate it back the opposite way, counting the number of clicks until it stops again (this may be any number of clicks between 200 and 600). Divide the number of clicks in half and rotate the knob in the original direction by that number of clicks. For example, if the total number of clicks is 500, rotate the knob back 250 clicks in the original direction. Center the elevation using the same procedure with the elevation control knob. The total amount of elevation clicks also varies between 200 and 600.

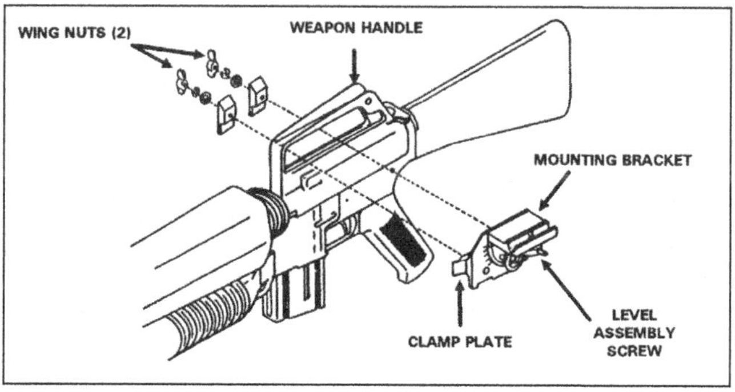

Figure 5-16. Installing the mounting bracket.

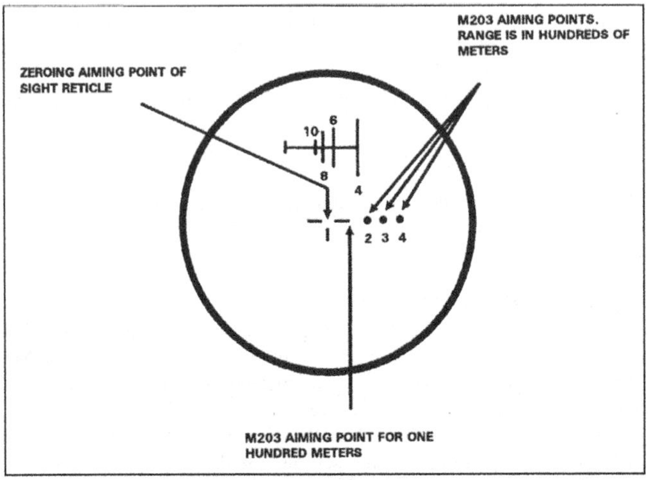

Figure 5-17. Aiming points.

b. **Adjust the Reticle Pattern**. Before adjusting the reticle pattern, the grenadier should fire three 5.56-mm rounds and then retighten the mount wing nuts to securely seat the sight. Once this is done, the grenadier fires at a 10-meter target because hitting and spotting this target is easier than hitting a 25-meter target. This procedure may be performed in daylight using the daylight cover:

(1) Turn the sight on and adjust the reticle intensity to the desired level of illumination.

(2) Place an M16 25-meter target at 10 meters and stabilize the weapon.

(3) Fire a 5.56-mm round at the center of the target and mark the hole the round makes.

(a) If the round misses the entire target, reseat the sight exactly as previously described.

(b) If the round hits the target but not within 20 centimeters (8 inches) of the center, adjust the azimuth and elevation controls to bring the impact point toward the center of the target. Then fire a second round. Continue to fire single rounds and adjust the controls until the rounds strike within the desired distance from the center.

(4) Once the reticle is adjusted, move the 25-meter target out to 25 meters and zero the grenade launcher. Do *not* remove the nightsight from the weapon until you have obtained a zero.

c. **Zero at 25 Meters**. This zero is not recorded. To obtain a 25-meter zero, the grenadier must—

(1) Stabilize the weapon.

(2) Center the reticle's zeroing range aiming point on the target aiming point, which is in the center of the target (Figure 5-18). Fire until you obtain a good three-round shot group. Triangulate and locate the center of the shot group.

NOTE: Even if the nightsight is dismounted and remounted on the same weapon, some changes in its zeroing will occur, so it must be zeroed again.

(3) Turn the azimuth and elevation control knobs to adjust the sight reticle. Move the center of the shot group 9.8 centimeters (3 7/8 inches) below and 4.2 centimeters (1 5/8 inches) to the right of the target aiming point (Figure 5-15). For example, if the shot group is high and to the left of the desired impact point, adjust the elevation down (DN) and the azimuth right (RT). One click of the azimuth or elevation adjustment moves the strike of the round .63 centimeter (1/4 inch) at a range of 25 meters. Two clicks move the reticle about one square on the target.

(4) After adjusting the reticle, assume a stable position. Place the reticle aiming point on the target aiming point and fire three more rounds.

(5) Repeat steps 3 and 4 until the rounds strike within a 3.2 centimeter (1 1/4 inch) circle in the desired location 9.8 centimeters (3 7/8 inches) below and 4.2 centimeters (1 5/8 inches) to the right of the aiming point, or until you have fired 12 rounds, whichever occurs first. If you are unable to zero the AN/PVS-4 after 12 rounds, the trainer must send you to remedial training.

(6) Confirm the zero on the grenade launcher range using a 200-meter target. Place the nightsight into operation and use its reticle, which has two parts. Use the vertical line in the upper part of the reticle to estimate range and the lower part to aim the weapon.

(a) Set the range as estimated on the range indicator of the mounting bracket (Figure 5-16).

(b) Engage the target, placing the aiming point of the sight reticle on the target's center of mass (Figure 5-18). Fire the weapon using all your marksmanship skills. You have confirmed the zero if two of three rounds strike within 5 meters of the target.

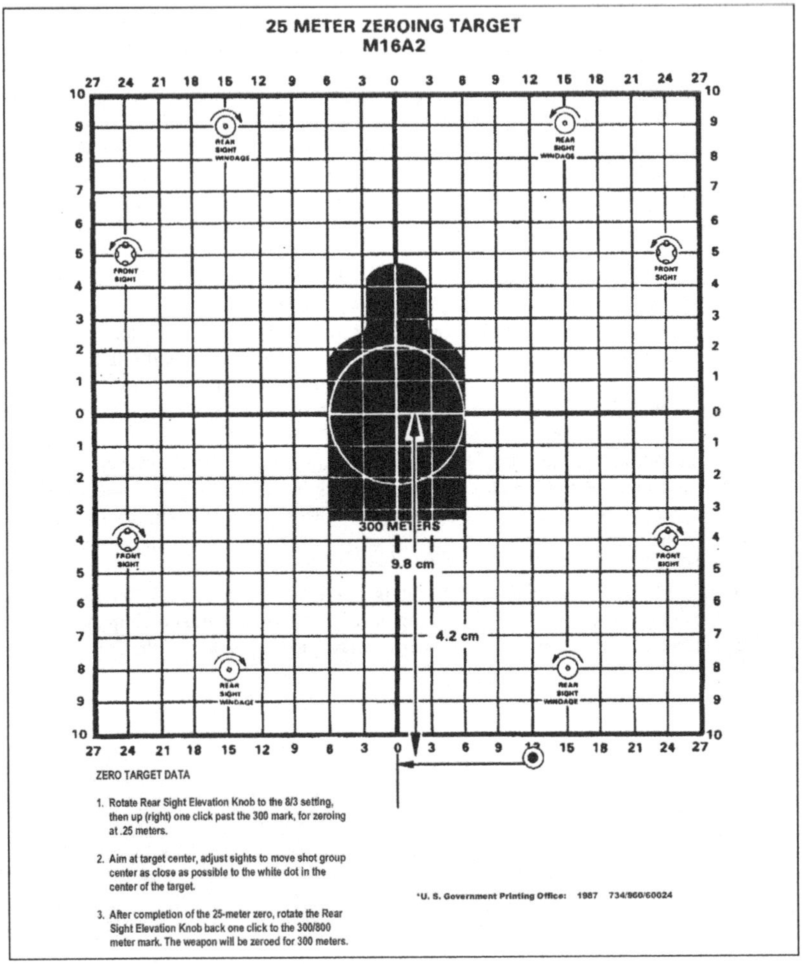

Figure 5-18. Adjustment of rounds.

5-15. NIGHT RECORD FIRE

Night or limited visibility firing trains grenadiers to apply the fundamentals of grenade launcher marksmanship while using the AN/PVS-4 nightsight. This training increases the grenadiers' confidence. Each grenadier learns to zero the M203 to the AN/PVS-4 on the 25-meter range and then fires at one area target on the M203 grenade launcher range. Before night firing, grenadiers receive instruction in its objectives, fundamentals, fire commands, and targets. The unit is organized in firing orders, each consisting of a grenadier and assistant, based on the range constraints. The assistant performs his duties in a manner similar to day record fire. Night record fire consists of one task, *Station 3, Task 6, Mid-Range Fighting Position, Area Target, Night*:

a. Load one of the three rounds allotted.

b. When you receive the following fire command, repeat each element as it is given:

GRENADIER
FRONT
3 ROUNDS
190 METERS (AUTOMATIC WEAPON POSITION)
AT MY COMMAND

c. Acquire the proper sight picture and announce "Up" to the grader.

d. When the grader gives the command FIRE, engage the target given in the fire command until you hit it. Fire any remaining rounds at the second target. You need no other fire command. For each round you fire, your assistant announces "Hit" or "Miss."

5-16. NIGHT RECORD FIRE QUALIFICATION STANDARDS

Before qualification firing, each grenadier must know the task, the time and ammunition required for each, the procedures to follow if a stoppage occurs, the penalties for failure to stop firing when commanded or signaled to do so, and the method used for scoring targets.

a. **Time and Ammunition**. Firing Table II (Table 5-2) provides the night firing task and its time and ammunition requirements.

b. **Stoppages**. The procedure for stoppages is the same as for other qualification firing exercises.

c. **Penalties**. The procedure for penalties is the same as for other qualification firing exercises.

d. **Target Scoring**. The target scoring procedure is the same as for other qualification firing exercises. Record the score in the appropriate block on DA Form 2946-R.

e. **Conditions**. Night record fire trains the grenadier to engage targets between 165 and 250 meters under ideal moonlight conditions.

STATION	TASK	TIME	ROUNDS	TYPE	TARGET(S) AND RANGE(S)
3	3	2 min	3	HE	Automatic weapon at 190 to 200 meters

Table 5-2. Firing Table II, night record fire qualification.

This page intentionally left blank.

(3) ***Linear Targets with Depth***. The grenadier lays on the target's center of mass. He then moves the grenade launcher left and right across the target, selecting successive aiming points at different ranges (Figure 6-9).

Figure 6-9. Linear targets with depth.

* 6-6. DECONTAMINATION

Leaders must try to reduce the penetration of contaminants and lessen exposure to them. Contaminated material is disposed of in accordance with the SOP. Refer to Chapter 2 for decontamination procedures.

Section II. FIRE CONTROL

Fire control includes all leader and Soldier actions in planning, preparing, and applying fire on a target. The leader selects and designates targets, indicating their width and depth or, in the case of targets that are hard to identify, designates the distance from a reference point to the target's center of mass (Figure 6-10, page 6-10). He also designates the midpoint, flanks, or ends of a target unless these locations are obvious to the grenadiers. The grenadiers open fire when ready, adjust and regulate the rate of fire, and shift from one target to another. They cease fire only when the target is neutralized or the leader signals to cease fire.

Figure 6-10. Use of a reference point to identify a target.

6-7. METHODS OF FIRE CONTROL

The noise and confusion of battle may limit the methods of fire control used, so the leader must select the method(s) that will best accomplish the mission.

a. **Oral Commands**. The primary method of fire control is the oral fire command. This method is effective unless noise or distance prevents the grenadier from hearing the leader.

b. **Arm-and-Hand Signals**. This method of fire control is effective only if the grenadiers know the standard arm-and-hand signals and can see the leader. The leader gets the grenadier's attention, then points to the target. When the grenadier returns the "ready" signal, the leader commands the grenadier to fire.

c. **Prearranged Signals**. This method of fire control can include visual or sound signals such as those that can be produced by a whistle, pyrotechnics, or casualty-producing device. The SOP must define the signals to be used, and all squad members must understand them. If the leader wants to shift fire at a certain time, he gives a prearranged signal such as smoke or pyrotechnics. When they see this signal, grenadiers shift their fire to a prearranged point.

d. **Personal Contact**. This method of fire control is the one most frequently used by leaders of small units. Many situations require the leader to move to individual Soldiers to issue orders. If so, he must use cover and concealment to avoid disclosing their positions. Once there, he gets the grenadier's attention, points out the new target, and commands FIRE.

e. **Standing Operating Procedure**. This method of fire control refers to actions executed without command. The SOP defines these actions and the events that initiate them. Using an SOP simplifies the leader's job of fire control.

f. **Range Cards**. This method of fire control requires the leader to ensure all range cards are current and accurate. He designates dead spaces, specific targets, no-fire zones, and restricted fire areas. The key to this method is the disciplined grenadier who pays attention to detail and can understand the areas the squad leader wants covered by fire. Figure 6-11 shows an example of an M203 range card.

*GLOSSARY

AAR	after-action review
ammo	ammunition
C	Centigrade
CBRN	chemical, biological, radiological, and nuclear
CCC	Captain's Career Course
CLP	cleaner, lubricant, preservative
cm	centimeter(s)
CMF	career management field
CS	a chemical agent (tear gas)
DA	Department of the Army
DODAC	Department of Defense Ammunition Code
EST	engagement skills trainer
F	Fahrenheit
FM	field manual
FOV	field of view
FPL	final protective line
fps	feet per second
FTX	field training exercise
HE	high explosive
HEDP	high-explosive dual-purpose
hr	hour(s)
IAW	in accordance with
IET	initial entry training
KD	known-distance (range)
kg	kilogram(s)
kps	kilometers per second
LFX	live-fire exercise
LTA	local training area
METL	mission-essential task list
m	meter(s)
min	minute(s)
mm	millimeter(s)
MOPP	mission-oriented protective posture
mph	mile(s) per hour

mps	meter(s) per second
MTA	major training area
MTP	mission training plan
NCO	noncommissioned officer
NCOES	Noncommissioned Officer Education System
NCOIC	noncommissioned officer in charge
NSN	national stock number
NVD	night vision device
OBC	Officer Basic Course
OIC	officer in charge (of)
OGIVE	head of the round
PD	point-detonating
PDF	principal direction of fire
PMT	preparatory marksmanship training
psi	pounds per square inch
RETS	remoted target system
rpm	revolution(s) per minute
S2	Intelligence Officer
S3	Operations and Training Officer
S4	Supply Officer
SEE	small emplacement excavator
SM	soldier's manual
SMCT	soldier's manual of common tasks
SOP	standing operating procedure
SSN	social security number
STP	soldier training publication
STX	situational training exercise
TP	training practice
TRADOC	Army Training and Doctrine Command
TRC	training readiness condition
UO	urban operations

*REFERENCES

SOURCES USED

These are the sources quoted or paraphrased in this publication. All are US Army publications unless stated otherwise.

FM 3-3	Chemical and Biological Contamination Avoidance. 16 November 1992, with Change 1, 29 September 1994.
FM 3-11.4	Multiservice TTP for NBC Protection. 2 June 2003.
FM 3-11.5	Multiservice TTP for Chemical, Biological, Radiological, and Nuclear Decontamination. 4 April 2006.
FM 21-60	Visual Signals. 30 September 1987.
FM 21-75	Combat Skills of the Soldier. 3 August 1984.
FM 3-22.9	Rifle Marksmanship, M16A1, M16A2/3, M16A4, and M4 Carbine. 24 April 2003, with Changes 1 through 3, 22 January 2004 through 28 April 2005.
STP 7-11B1-SM-TG	Soldier's Manual and Trainer's Guide, MOS 11B, Infantry, Skill Level 1. 28 May 2004.
STP 7-11B24-SM-TG	Soldier's Manual and Trainer's Guide, MOS 11B, Infantry, Skill Levels 2, 3, and 4. 28 May 2004.
STP 21-1-SMCT	Soldier's Manual of Common Tasks, Skill Level 1. 11 October 2005.
TC 25-8	Training Ranges, 25 February 1992, with Change 1, 5 April 2005.
TM 9-1010-221-10	Operator's Manual for 40-mm Grenade Launcher, M203. 1 August 2001.
TM 9-1010-221-23&P	Unit and Direct Support Maintenance Manual, Including Repair Parts and Special Tools List, for Launcher, Grenade, 40-mm, M203. 12 January 1993.
TM 43-0001-28	Army Ammunition Data Sheets for Artillery Ammunition: Guns, Howitzers, Mortars, Recoilless Rifles, Grenade Launchers, and Artillery Fuzes. 28 April 1994 with Changes 1 through 13, 30 May 1995 through 31 March 2005.

TM 750-244-7	Procedures for Destruction of Equipment In Federal Supply Classifications 1000, 1005, 1010, 1015, 1020, 1025, 1030, 1055, 1090, and 1095 To Prevent Enemy Use. 18 June 1970, with Change 1, 29 October 1970.

DOCUMENTS NEEDED

These documents must be available to the intended users of this publication.

AR 385-63	Policies and Procedures for Firing Ammunition for Training, Target Practice, and Combat. 19 May 2003.
DA Form 2946-R	40-mm Grenade Launcher Scorecard.
DA Pam 350-38	Standards in Weapons Training. April 2006.
TRADOC Reg 350-6	Enlisted Initial Entry Training (IET) Policies and Administration. 30 December 2005.

WEB SITES

Army Knowledge Online	https://akocomm.us.army.mil/usapa/doctrine/index.html
Army Publishing Directorate (for forms)	http://www.apd.army.mil/usapa_PUB_formrange_f.asp
Reimer Digital Library	http://www.train.army.mil

*INDEX

40-MM GRENADE LAUNCHER SCORECARD

For use of this form, see FM 3-22.31; the proponent agency is TRADOC.

1. ID CODE *		2. UNIT	3. DATE *(YYYYMMDD)*
4. ZERO LEAF SIGHT	DEFILADE	ELEVATION	
5. ZERO QUADRANT SIGHT	DEFILADE	ELEVATION	

6. RECORD FIRE	TASK NO.	TIME (MIN)	TGT	HIT	MISS
DAY	1	2	1		
			2		
	2	2	1		
			2		
	3	2	1		
			2		
CBRN	4	2	1		
	5	2	1		
NIGHT	6	2	1		
* Do not use personal information	7. TOTAL POINTS *(Award 10 points for each hit)*				

8. RATING SCALE

TOTAL POINTS *(Block 7)*	RATING	
80 - 90	EXPERT	☐
70 - 75	FIRST CLASS	☐
60 - 65	SECOND CLASS	☐
0 - 55	UNQUALIFIED	☐

9. GRADER'S INITIALS	10. DATE *(YYYYMMDD)*
11. OIC'S INITIALS	12. DATE *(YYYYMMDD)*

DA FORM 2946-R, FEB 2007 EDITION OF NOV 2002 IS OBSOLETE. APD V1.00

Made in the USA
Monee, IL
05 November 2026

67808337R00177